'Born Lucky'
Book Two

From Mynah to Miner and Beyond

By Donald Povey

Copyright © 2023 Donald Povey

All rights reserved. No part of this publication may be reproduced, distributed, or transmitted in any form or by any means, including photocopying, recording, or other electronic or mechanical methods, without the prior written permission of the publisher, except in the case of brief quotations embodied in critical reviews and certain other non-commercial uses permitted by copyright law. The moral rights of the author have been asserted.

I was inspired to write this second book by the people that had congratulated me on the first one and had enquired if there was a second forthcoming. As I am a person not used to saying anything negative, it started me thinking of more experiences and friends that I'd made that had really enriched my life. The more I think about it, money is fine for perceived security, but friends and relatives give something beyond that, far beyond that. Therefore, I make no apology for mentioning lots of names again, they may mean nothing to you but mean a lifetime to me.

Thank You

Ibstock Historical Society
Coal Mining Equipment UK
Coalville Historical Society
South Midlands Coal Mining Heritage Group
Coal Mining in all Areas
The Pits
Coalville Mining Memorabilia
World Wide Deep Mining
Spotted Ellistown
Adrian Brant
Mick Westwood

for all been so very helpful and obliging.

"Trust" by Mick Westwood

My life was as a miner, and that is where I found,
The greatest of companions, while working underground,
Each day presented challenges, you had to overcome,
To miners that was natural, or work became your tomb.

We worked on `trust` together, and `trust` became a chain,
He watched me, and I watched him, no - one could abstain,
And in a coal - mine's dim lit glow, where all was grim and black,
You had that reassurance, that your mate would,
`watch your back`.

It didn't matter where you worked, in a team of two, or ten,
This `law` although unwritten, was observed by mining men,
This `brotherhood` extended out, beyond the colliery gates,
To enfold all of the family, not just the miner and his mates

And this is why I'm proud of them, the miners of this land,
Because, some of them, at some time, held my future in their hand,
And safe in their entrustment, they did not let it go,
They returned it to me safely, - that's why I loved them so.

'Born Lucky' Book Two: From Mynah to Miner and Beyond

Meadow Row

Meadow Row was where I was born, originally built as miner's homes for the local coal mine that was less than a mile away with a brick and pipe yard between and close to the mine. I remember very clearly what it was like, a row of sixteen houses with an 'entry' between numbers eight and nine and all in a straight line with a causeway (corsey) along the front of the terraces for the rows whole length. Also at the front of the houses was the garden that reached to the Spring Road, also known at the time as 'Pit Lane' as it was unsurfaced and used as a walkway to the Ibstock coal mine which closed in the early 1930's. At the rear of the 3 bedroomed houses was a 'black pad' that was some 10 metres wide and had to be crossed to get to the outside toilet, coal house and a storage area, I had no idea what that was intended to be but it was a play area for us. That black pad stretched the length of the meadow as we termed it and was used for access for dustbin collections, deliveries and cars etc.

'Born Lucky' Book Two: From Mynah to Miner and Beyond

Me at age 5

Meadow Row as remembered by Danny Povey

Meadow Row

Beyond the outbuildings were Dolman's pig farm, Black's field which had a few caravan homes on it and a footpath leading to Chapel Street and local shops, pubs and clubs. The ground behind the Meadow sloped from Chapel Street to the Spring Road with the Meadow being between the two. In times of excessive rainfall water collected and ran from the field through the outbuildings and would flood the houses as the drains became blocked. I remember a couple of times Mum and Dad having to raise cupboards, sideboards and settees etc. onto house bricks clear of the floor to protect them and take up rugs to dry them out. The brook that flowed along Spring Road became swollen, overflowed and spread across the full width of the road and running over Curzon Street/Pretoria Road (luckily no houses at that point) and then rejoining the brook at Jack Newbold's field (Local farmer). There used to be a wooden bridge on the far side of the road for pedestrians to pass over the flooded road to get to Pretoria Road (common hill) and to the cemetery.

The first recollection I have is a Christmas time, I was given a three wheel bike with a box at the back and a small basket at the front, also a bell. I was so excited that I rode it all day only going in home when Mum called me to have something to eat. I rode in a figure eight along the first half of the corsey, through the entry and along the black pad around the end of the houses, back along the second half of the corsey, through the entry again and completed the circuit by doing the second half of the black pad. Mum called me in for the last time as it fell dark. I complained that I could still see by the light from the house windows but when I was in home I realised how cold I had become and was glad to see that glowing coal fire, I was not yet school age.

Brother Roger came along as did our sister Kathleen (Kate) the other kids along the Row made a nice little group for playing outdoors with, there was Pete Rowley, Norman and Billy Gould, Gerald Taylor, Roger and I, Ivor Gray. The girls were my aunty Sue (younger than me) Kate, Helen and Sheila Tomlinson, Pearl and Peg Tomlinson, Carol Ball, Josephine Gould, Jacqueline and Josephine Finney, in fact anyone that was of a similar age. There were of course some other kids but either too young or too old for us to play with. Other friends of similar age came to play with us on a disused part of the Miner's Welfare next to the pit lane (Spring Road) brook, they would come from common hill, Leicester Road, Curzon Street and Chapel Street for self-organised games of cricket, football, rounders, jumping the Brook, fishing, blackberrying, flower picking and any other harmless pass-times we could invent.

With being so close to the brickyard it was inevitable that we would stray there through curiosity, no one ever questioned us or sent us off the property, no boundary fences were in place nor any sort of security. We would build dens in the haystacks (straw was used to spread between layers of bricks as they hand stacked from the kilns onto lorries to minimise breakages). We would play on the hilly areas on the far side of the works that were presumably formed from the dumping of the original overburden to expose the proper clay beneath. Those small hills were a favourite for playing cowboys and Indians also cops and robbers but were excellent for use as ski slopes when wet. What I can only describe as small pallets, formed from three strips of wood parallel to each other connected by steel rods, the whole thing measuring one metre by half a metre, were utilised as sledges or snowboards. These small hills were grey in colour and our play clothes got filthy from constantly falling and sliding and very wet at times from the puddles that formed in the valleys between the hills. No problem though, before going home we would go to the circular brick kiln area and stand close to the several open coal fires around the periphery of the kilns and dry ourselves off.

A severe telling off from our parents was expected and delivered with great anger from our mothers, often with the threat of a beating with the 'copper stick'. In those days every house had a copper, often in a corner, brick built with an open topped container (the copper), fitted with a thick heavy wooden lid and with a coal fire underneath the copper. This was used for a boil wash, of workwear and any other garments or bedclothes that warranted boiling would be dealt with here. When the wash was boiling nicely a stick was used to agitate and turn the wash, that stick became bleached white with use, and being about 0.75m long made a handy weapon and was used as such around our legs when we deserved it. Ouch!, I still remember it, the threat was often enough!.

Looking back, the most frightening thing we did as kids was to slide on the frozen tops of the clay holes' water. The sides of these cavernous holes were very steep but on our nearest one, the 'old clay hole' there was a side that was unused. Trees grew on that part enabling us to access the frozen water by holding on to the trees and branches, lowering ourselves until we reached our destination.. There would be six or seven of us girls and boys, maybe in the age range of eight to ten. On reaching the ice the first thing to do was test the edge nearest the shore, if that passed the test the next thing was to throw large stones or part bricks on to its surface, thus by listening to the sound that the missiles made on contact with the ice gave us an estimation as to its thickness. Then followed a discussion as to whether it would support our weight. After deciding that 'maybe' it would, the next thing would be as to where someone would make the first tentative boarding of the frozen surface. This was deemed to be a small atoll some ten metres towards the centre of this fifty metre or so wide lake. Now, who was to make this crossing? A debate would take place with points being made like 'My Mum will kill me if I get wet', 'I'm the youngest', 'I'm too heavy'. All the excuses that could be thought of. Until someone would come out with 'I dare you', this was big talk at this point of the negotiation, one of us lads would be offended and there would be a period of mental preparation as the miffed party made the necessary calculation as to how he could make that perilous journey. Alas, at long last that journey would be made, a report shouted back as to what it felt like and was it safe for another to cross or did the ice crack. Others would cross until the atoll could hold no more; it was now time to move further. Of course the initial pioneer said, 'I've done it once, it's not my turn!' The others would make their excuses and the pioneer, buoyed by his success, would make the next leg of the expedition. So it went on, until enough confidence was gained to run and slide across the middle of our lake.

It was during such an escapade, having gained the confidence on the ice that a clump of bull rushes caught my eye near to the 'live' side of the clay hole. I ventured to them as I had never held a complete bull rush reed, they had always been opened and shedding their seeds. This was too good to miss, so I looked and felt the nearest one, then the biggest one, I wanted to show my friends so I got as good a grip as any nine-year-old could and pulled with all my might. I thought the reed was coming out of its bed, it wasn't it was the ice breaking and I fell through and was up to my chest in ice cold water. Try as I might to pull myself onto the ice, it kept breaking away until at last I used my elbows to spread my weight and managed to clamber onto the ice and to relative safety.

I didn't shout for help although the group had made their way to me, I was drenched and freezing cold, my only thought was to get dry, mum and dad would go spare. We made our way to the brick kilns up the bare steep side of the clay hole to the roaring fires that we knew would be ablaze and I dried myself until I thought it would be acceptable and no one would notice. How wrong I was, dad forced out of me what had happened and gave me a stern talking to, mum was crying as to what might have been and I was left in no doubt as to not do it again, Roger and Kate were also warned.

It was shortly after that that news came of a local lad, older than me, had been drowned in that clay hole after the ice had gone. Born lucky or what!

Banned from those clay holes we continued to play footy on the 'welly' (miner's welfare) until, getting bored again we ventured back to the hilly lands near the brickyard, by now it was summer and the old style of play re-enacted, cops and robbers etc. with full sound effects and arguments as who'd been shot and was out as the game. That was old hat and we were looking to expand our horizons. That attitude took us to closed Ibstock colliery site and the reservoir, or the 'resser' as we knew it. Dad, Roger and I would carry course freshwater fish in buckets and stock the resser with them in years to come and it was adopted by Ibstock Angling Club for a time.

The headstock and buildings were long gone, all that remained of interest to us were two brick-built domes about ten feet in diameter and around six feet high, probably the mine shaft caps after they had been filled in.

What remained was big zinc doors, a mine vehicle (tub) on a small rail track that led to the zinc doors. What could it be for?. Getting up to the zinc doors and spying through chinks and gaps we could see an arch shaped tunnel going downhill as far as we could see accompanied by the rail track and suspended pipework, that was all that was to be seen, boring really. The pipework extended outwards to the daylight and into a big tank some ten metres long by five metres wide and at one point five metres high, a settling tank for the old mine water, which overflowed by intention to a water course and away. That short tunnel, I found out later was termed 'The Footrill' and was kept open to control water levels at nearby South Leicester Colliery and was examined statutorily by their officials. Even further afield were the train tracks and on occasion we would see a loco being bought from its shed somewhere near the top of the bridle path between Pretoria Road Ibstock and Whitehill Road Ellistown.

Halcyon days, carefree, free to wander wherever, explore, question, wonder, touch and feel anything we thought wouldn't get us into trouble and even if it did, how could we avoid getting found out?

One thing we never tired of was following a brook, trouble was that they passed through farmer's fields and in those days they would shout and wave their fists at us making us take flight to wherever we felt safe, which was often an abandoned barn or other farm building, copse or hedgerow. Our local brook on Spring Road or Pit Lane as we knew it was our favourite for fishing, building dams (stank) across it until they burst, making rope swings over and paddling in, in summer. There was a wooden bridge over it near a big tree, one of whose branches was perfect to tie a rope to and swing from bank to bank. If only someone was brave enough to climb the tree to attach the rope!. Next to that tree was a favourable height hedge of Elder and Hawthorn bushes. As the big tree had no lower branches the only way was to scale the prickly Hawthorn onto the Elder and then spring to the first branch of the tree and pull yourself up using your arms and wrapping your legs around the branch. That in itself presented a problem, you were then hanging below the branch like a koala bear, but determination, tenacity and shear doggedness would get you on the upper side of the branch. The rest was comparably easy, I loved climbing that tree, if you got high up no one knew that you were there and you could people watch over the cemetery, Spring Houses. Meadow Row and the Miners Welfare going about their daily business.

The Spring houses were named as such because of the spring water that surfaced there, on the footpath side of the brook next to where it was piped underground on the upstream side. A second spring could be found on the same brook downstream somewhere near the bottom of Grange Road and Orchard Street.

Once the rope was attached, (we called that strength of rope, a tow rope) it hung centrally over the brook. Kids came from the locality, and we made new friends such as Chris Statham, Pete Findley, John Whitby, Josephine and Jacqueline Finney, and Rob Storer. The kids awaiting their turn on the swing would be on the wooden bridge watching and daring the kid swinging to let go of the rope and try to make land on the far bank, this always ended in tragedy, the performer nearly always falling in the brook, crying and exclaiming 'my mum will kill me'. We knew that wasn't true, but it would be the dreaded copper stick!

All this period of time mum worked full time in Ward's shoe factory at Barwell and Grandma Ibstock (Flo) looked after us from an early age. My aunty Sue was akin to a sister to us, and still is, as she was the same age as me and the four of us spent a lot of time together playing on Ibstock Penistone Rovers football ground, immediately behind Gran's house on Central Avenue. Occasionally our cousins would join us, Kay, Marilyn and Hazel Harrison, Yvonne was older and was past such triviality, her being the role model for us all!. I remember being trusted to walk to and from the infant's school on Grange Road to her house unaccompanied at age six, there being very little traffic in those days and the occasional horse and cart delivering/selling goods and wares.

So it continued, those happy days through Infants and Junior School and on to Secondary Modern school, virtually across the road from Gran's. I was at the 'Senior School' because I failed the eleven plus exam, although being heavily tipped to pass it. I recall how important it was to our family because there was a meeting with Gran, Granddad, Mum, Dad, Aunties, Ann, Sue Marion, Winnie, Kath and Mary. All promising money if I passed totalling around thirty pounds I remember, a good sum in those days. But alas, not to be, I was to be a failure and become a duffer. Not if I had anything to do with I wasn't!.

So, out of short trousers and into long ones, I was now moving up in the world at last. Senior school was great. I was put into the 'B' stream, the streams being A, B, C, and D. I did reasonably well finishing in the top ten but not well enough to get promoted to the A stream.

In my third year in a gym lesson, first period after lunch in the hall, I was scaling the climbing frame, got to the top and fell off, landing heavily on the wooden floor. I remember big Mr Smith (There were two Mr Smith's and the whole school, teachers included referred to them as big and little Mr Smith's) picking me up, Mr Cross (Ex-navy, broad chested with an awesome loud voice at times, everyone respected and liked him) coming into the Hall and asking what had happened, Mr Smith replying "He fell, look at his right wrist". With that I looked at my wrist too, it was a strange shape, there was a distortion where my wrist had been. With his arms under mine, Mr Smith virtually carried me out of the hall, up the stairs leading to Mr Warner's office (Headmaster) and the first aid room, my feet didn't touch those steps. "I bet you think this is the end of the world Povey, but it's not. You'll be right as rain in no time at all". Mr Cross assured me, gave me a drink of water, put my arm in a sling and got me to his car. Very quickly we were at Ibstock doctors' surgery, straight in to see Dr. Meldrum who examined me, fitted me with padded splints and a new arm sling and phoned for an ambulance.

Mr Cross waited with me until the ambulance arrived and left after he'd seen me safely into the ambulance and being whisked off to Leicester Royal Infirmary. I learned later that the school and the hospital could not contact anyone to get permission to give me anaesthetics to treat my injury, no one had telephones at that time, Mum and Dad were working and Gran wasn't at home. The school was very close to my Gran's on Central Avenue and someone from the school had the forethought to leave a note asking her to contact them. That done, the system took until six in the evening to get the granted permission to filter through to the doctors at the Royal. After I had been treated, I was ferried home by ambulance arriving at nine o'clock at night to a tearful Mum and a comforting Dad and sporting a plaster cast from shoulder to wrist, covering my knuckles and encircling my thumb. After the pain had subsided I was very proud of my plaster.

I was off school for six weeks and it was summertime, not sure if the school complained but if they did I didn't know of it. Those weeks were spent mainly with my dad, walking in the fields, footpaths and lanes with him telling me all about the nature he loved so much; about rabbits and their habitat and how he and his dad would catch them for the table, knitting their own 'posh' and 'long' nets to catch their prey with a system they called 'Pegging'. Pointing out different birds, in particular Goldfinches, linnets, wrens and his favourite Bullfinches, going on to explain their habits, birdsong, breeding and feeding differences. Dad loved to breed Yorkshire canaries and was well known amongst local enthusiasts and also abroad, mainly Belgium, Holland and France. The more he won competitions, the more his stock was worth, and breeders would visit his aviary and purchase some of his birds. He made his own cages to the required standard and would give any excess to various societies or to novices just setting up. All this popularity had its price as one night after his retirement he was robbed of all his best birds and so he gave up his favourite hobby. Those few weeks made me appreciate my dad more than ever and we were closer than we'd ever been.

At aged thirteen we left the Meadow; the houses had been condemned and we moved a good fifty metres to Spring Road No.17. These houses were built at the bottom of the gardens of the old Meadow Row, what a revelation! A downstairs and upstairs toilet, bathroom, hot and cold water, bigger bedrooms, new kitchen and utility room and a lounge big enough for two three-piece suites, all brand new. Neighbours were virtually the same except now I had Josephine and Jacqueline next door, finding out later that they were dancing in their kitchen to my recorded pop music that I played loudly as I bathed on Saturday afternoons ready going out on Saturday night. That often led in later years to party time at the Finney's on Saturday nights, wow! They were some parties.

The third and fourth years at school were amazing, I was part of a lot of activities, mainly sport, rugby, cricket, football, swimming and badminton after school with Miss Stiles. I was also a member of the chess group. Representing the school at rugby, swimming and diving. Richard Cockle, Ernie Dinsdale and me forming our own trio. Ben Clinton, Johnny Saunt and Pete Brown were another clique, other lads seemed to drift between our two groups as it suited them. At last I made to the A stream via an unusual route. Cockle and I had gone from

3B to 5A to have a crack at the GCE grades, it didn't work out and we dropped out and were placed in 4A. I didn't look back, Dinsdale and I became milk monitors, always drinking two or three bottles at the crate stack before taking the required amount to our class. Also Mr Cross had handpicked us to take the first years, every Thursday morning for swimming lessons at Coalville Baths on the hired coach, great stuff!. We would take one coach load, Mr Cross would do their lesson, we would make our way back on the empty coach and pick up the next classes, when we got back to the baths we would re-load the coach with the first classes of youngsters, feeling very superior all the time. We would hurry back to the baths and get our trunks on and join the second class in the pool, encouraged by 'Crossy' who had been in the pool all the time, he loved it!.

At about that time I began to notice that girls were different to boys, and I started to take an interest in them. They weren't just useless at climbing trees, football and didn't want to play cricket or rugby but something made me want to associate with them as well as the boys, whatever could have changed?. I would go up Common Hill, as we knew Pretoria Road, and meet with Chris Statham, Terry Lawrence and others because I knew a lot of girls up there from school, Marion Ball, Sue Tompkin, Sandra Gray, Sonia Astley, Christine Bott, Margaret Truman to name a few. Still playing the usual games like hide and seek, tin tin a-lurkey and snobs under the streetlights but somehow it was more fun. In the summer I would cycle to Newbold Verdon on my own to see girls from my class like Janet Bills, Val Satchwell, Jean Devereux, Janet Collett and others also meeting up with Melv 'Tucky' Tuckwell, Ian Mitchelson, Cockle. Dinsdale and Bob Hunt, but it was the girls that made the trip worthwhile, a bonus on the way back through Bagworth was if I bumped into Sandra Spencer and/or Patricia Bateman. What a variety as friends I had.

Alas those days were racing by but my interest in girls wasn't waning, one lunchtime a group of us were in 4A classroom and Janet Collett and I were in close proximity to each other and some of the lads had told me that she liked the look of me and egged me on to kiss her. Never one to shirk a challenge, I moved in, and we kissed, it was blissful for a split second. "POVEY, STOP THAT AT ONCE!". Immediately. "No Sir, I like it!" I replied in my best manly voice. Mr Johnson had seen us through the window from the playing field and was observed running indoors, jacket flying, to separate us, all too late, the moment was spoiled. I got a telling off and a warning, but Janet and I were now a little closer. She even chose to sit on my knee at our last school sports day, born lucky or what?.

Best time of all was Friday afternoon, that was when the girls had 'Domestic Science' from lunchtime to end of the day.

They would be baking all of that time and the smell throughout the classrooms would be wonderful, all the boys being like 'Bisto kids'. When the final bell went it was every man for himself!. All the girls leaving for their respective buses and the Ibstock girls, were surrounded by boys begging cakes or whatever had been baked but apart from the feeding frenzy it also told the boys which girls liked the cut of your jib.

The school was being extended and modernised and a new domestic science building was built on, our girls in 4A were the first to use it, they had been set the challenge of cooking a three-course meal for the boys of our class. Such excitement!. It was the talk of our class all week. Then came the day, excitement turned to trepidation and anxiousness and that was just us boys!. We needn't have worried it went like clockwork under the supervision as Mrs Jones, Domestic Science teacher and Miss Harratt our form teacher, and the food was excellent, every boy being ravenous because normal lunch had been missed.

Then came the end of year and our final exam time, every subject had to be taken and results posted on the classroom wall as the subject was marked, ten subjects in all. This caused much excitement as to whom had been top of the class in that subject and as that became apparent whom would be top in the next subject and so on. The day came when all results were in and Miss Harratt (I later found that her name was Mabel, but you daren't take the mickey.) sat in front of the class and went through the class in alphabetical order, reading out their individual marks. We were tasked with writing down our own marks and totting them up for our final score. After that little exercise she read out our class position and I was satisfied with sixth place, considering the talent that our class had in it, boys and girls. The next morning after assembly Miss Harratt called the register as normal and said that she had checked the individual marks, asked for me to stand up (I was on the back row) and said "Povey, how on earth did you come top in mathematics and can't add up?" "Don't know Miss." I replied in the lowest tone I could muster to appear more manly. "You were one hundred marks astray and that one hundred has moved you from sixth to top as the class, sit down young man." "Yes Miss, thank you Miss". Don't know why I was so excited, I hadn't thought I would be top of my class at any point, I was only top in two subjects and that was woodwork and maths!.

After exams, school was all trips, walks, sports and generally letting off steam, within reason. A trip to the Shakespeare theatre at Stratford on Avon, walks around Great Glen (Leicestershire), sports day and being flippant with teachers, after all we were almost wage earners now and the world was at our feet.

Being Your Own Man

I was ready to start work, but there was unbounded fun to be had whilst I made up my mind where. What to do though, cowboys and Indians was old hat as was hide and seek, there had to be something else surely. Getting a new bike was the answer, and having got a job in mining, I was free to do as I pleased and with transport I could enjoy myself anywhere.

Hugglescote became our first go to place in those early days as there were local lads we knew and girls were about all the time, much better than the local town of Coalville we thought in this transitional period. Colin and Graham Nutting became good friends of ours and Norman, Pete, Chris, Barrie and I were regularly down there. Unfortunately so were a group of Ellistown lads, Bill White and Aubrey Rudin amongst them. This always started off well on meeting but at some point would flare up, the girls would scream and shout and us lads would be squaring up to each other, huffing and puffing, always coming to blows and wrestling for supremacy. Local adults would complain and try to mediate and this would almost always calm the whole lot of us down and we would be friends again until next time. Being a teenager is joyful but can be hard.

My uncle Barry said to me once "Were you fighting in Hugglescote last night?" I grinned at him and asked why. "One of my workmates had a swollen cheek and eye and I asked where he'd got it. He told me and mentioned your name. I told him I was your uncle." "You haven't told Dad have you?" I asked. "Of course I have," he answered with a knowing, wry grin. Mum carried on at me, but Dad asked later "Did you win?" "Yes, but my knuckles are sore," I replied. He was happy.

Good times (1964)

The Billiard Hall at Ibstock was a very regular haunt for us teenagers and occupied us for entertainment, competition and general high spirits. Believe it or not, the guy that ran it was called Ralph Cannon! We would spend a lot of time there. Entrance was via a door at the back of the building and up some semi-lit stairs as it was above a store called Worthington's at the time. On entering the hall you were greeted by a darkened room, narrower at the entrance end and widening out to the far end with two snooker tables, the nearest running along the hall and the furthest across the width of it. There were windows on the far and right hand side wall, all with drawn curtains. These were very handy for posting lookouts spotting Ralph coming back from the Whimsey or the local policeman that sometimes called in.

The lookouts were essential as we fiddled free plays from the juke box and the pin table, very naughty of us but only to be expected from a group of teenagers left to their own devices I thought!. The floor was of hard-wearing conveyor belt, around the outside on three sides was a wooden plinth of such dimensions that up-cycled bus seats were placed on it and used as viewing areas of the games being played. Along the shortest wall was a windowed alcove in which was placed a juke box and immediately to the left and up four steps was Ralph's command centre room where tables were booked (6d per half hour and 9d per hour, (2.5p and 4p) personal cues were hung and purchases could be made of chocolate bars and the like.

The only places that were well lit were the two tables, so it was difficult to see who was in the hall around the outside and the players only became visible when taking their next shot, which didn't take long as no big breaks were ever scored there, certainly not in our circle. Various forms of snooker and billiards were played, snooker plus being one as my favourites whereby two extra balls were bought into play, purple and orange. The purple ball being spotted between the pink and blue balls and being worth ten points, the orange ball being worth eight points and spotted between the blue and brown balls. Normal snooker rules applied but now much higher breaks could be attained.

It was during one such frame between Pete Rowley and I that I went in off the purple, giving ten points away, not being best pleased (There was a Mars bar bet on the game) I picked the cue ball from the pocket, threw it towards Pete at the other end of the table, bouncing it off the slate bed of the table. Pete made no attempt to catch it, I don't know if he even saw it!. It bounced off the table, slightly clipping the light shade, bisecting two lads looking at the juke box, over the juke box and smashing through the window behind it and into the dark night beyond. Everyone was stunned, I blamed Pete for not diving to catch it, he blamed me, the two lads didn't know what happened, Norman, Barrie, Chris and Roger were shouting "Find the cue ball, we're on next." Cue ball found in the yard outside, curtains drawn over the broken window and Ralph wouldn't know any different for a while, he was in the Whimsey, the pub next door as he often was mid-evening. I plucked up courage to own up to it and he let me off providing that I helped him clean the hall on Sunday morning. I did that and carried it on for a while. Ralph ironing the tables after I'd brushed them and then I'd sweep the floor and we'd have a frame of snooker before he'd lock up and go for a pint at the 'Whim'.

It was in the billiard hall one evening, shortly after Chris, Mick Brooks and I had secured employment at the NCB as trainees that we heard the news that John 'Hunk' Horner had been seriously injured at South Leicester Colliery and wasn't expected to survive. Luckily he did and went on to lead a normal life with a family of his own. It made us realise what we could be in for if we weren't very careful in our future work lives.

Working at Archie Dawson's farm on Chapel Street Ibstock alongside Norman I had money of my own in my pocket. We'd all sorts of jobs in the evening, I didn't attend in the morning. We'd fetch the cows home (6) and milk them, muck out and then, collect eggs from the hens that laid anywhere on the farmyard and little jobs that 'Arch' wanted doing. My favourite was travelling on the tractor and trailer to the 'Reccy' where Arch had some land near the allotments and go kale and mangold cutting with big machete's. The kale being around 1.2m high with big leaves that held lots of water, it was for cattle winter feed. Arch wanted the kale cutting at ground level, so we were bent double under the leaves hacking at the plant base, shaking it violently, we were soaked through after a couple of plants each but we didn't care, it was great fun we thought. Then on to the mangold's, these were fun too, pull them out of the ground, hack off the tops and roots with one blow each with the machete's and chuck them in the trailer, great stuff, not sure I'd think so now though!

With this new found wealth, now was the time to go along to Coalville and see the night life, as we thought. This consisted for us of riding around exploring all the alleyways, jitty's and public footpaths whilst daring to ride along them on our bikes, the height of our rebellion at that time. A severe telling off from a policeman was the penalty if caught along with the threat to 'tell your Dad' if you were known. Soon we were confident enough to go into the cafes of the town like the Rex cinema cafe, The Hacienda, Chad's, The Jay Bee and the the newly opened Wimpey Bar in the precinct, this was more like it. Local fairs were a big attraction and we would try all the big rides and the pellet guns and using the big rides structure as apparatus we would vault over the highest point and drop to the ground, hopefully unhurt

One particular Saturday night coming from Wileman's chip shop on 'Margo' (Margaret Street) passing the Rex Cinema Coalville and in high spirits I saw two girls standing on the top step, there were four, I bounded up the steps and said to them "Where are we going tonight girls?". They looked at each other and giggled. "I know, we'll take you to the

Wimpey bar." "Where's that?" they asked. After a bit of chat and laughter it turned out they were killing time until their bus came for Ashby-de-la-Zouch, so we had an hour for Barrie and I to make progress, he had approached the other one. They thought the Wimpey bar was very 'Posh' and didn't really want to go in, but we made out we were regulars and confidently took them in and treated them to a milk shake. Classy or what?. After escorting them to the bus stop we made a date for the following week at the Rex. That was my first encounter with my future wife, Sue Ison. I look back at this point in my life and feel that it was when I picked the path I wanted to take to make an independent life for myself, I was fifteen. It wasn't a smooth courtship by any means, we argued and fought, split for weeks on end but ended up back together.

Credit: Adrian Brant

Things were going well for me at the mine, I had done my training, got a good name and got a good pay rise. For that pay rise I was, as well as others, expected to do things lots of men wouldn't. One such task was to keep 80's face supplied with timber and steel. 80's face had just been installed and was a bit rough, it was equipped with roof-master chocks and the machine was a trepanner.

Pony

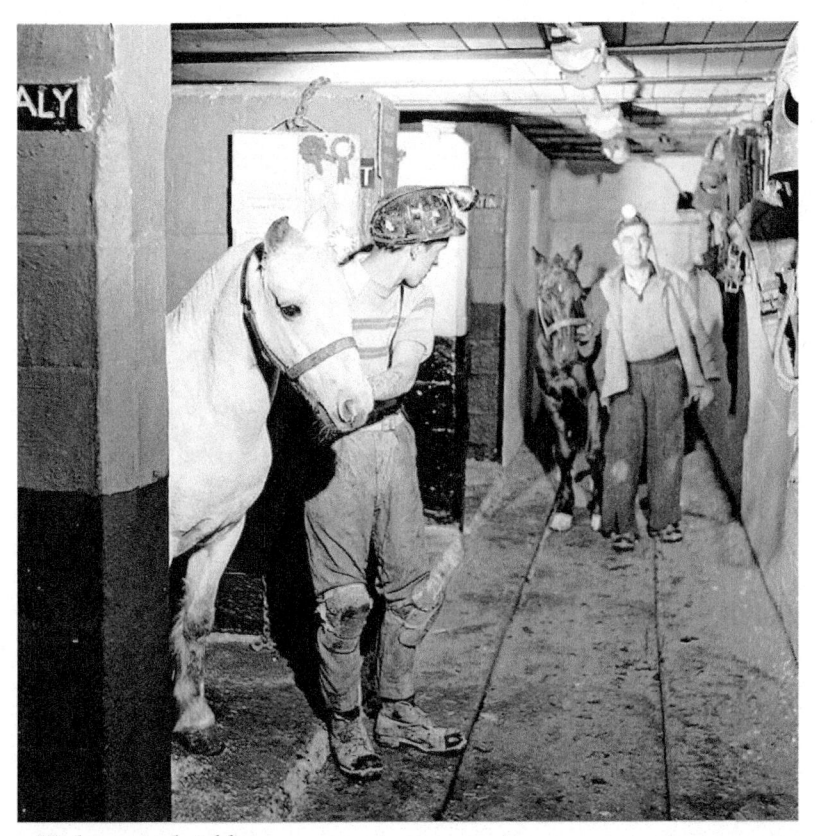

Underground stables

Hughie McDermott and I were tasked with getting supplies along the five-foot seam level by pony haulage, to the bottom as 80's drift (Underground hill), transferring to a direct haulage to travel the drift. That drift was horrible in that it was wet throughout its length, about 100m and a gradient of 25% rising (1:4) also at its mid-point a Minge seam roadway had thurled (connected) to it and this had caused mother nature to throw her considerable weight onto the drift and cause it to converge so much as to be impassable by a loaded mine vehicle. So Hughie and I having got our supplies to the base of the drift had to take them one load at a time up it. Ronnie Wardle drove the direct haulage and got us to the low area where we unloaded the supplies at the side of the road and raised the empty tram away from the area. The drift was wet under foot and also in parts dripping water from its roof, this made for bad working and walking conditions and Hughie and I began hand balling the supplies from our unloading spot through this low, wet area, one piece at a time and passing it to each other until the full load was on the uphill side of the wet and low area. Ronnie on the haulage awaited our signals and slowly raised the empty tram on its rail track to a point where it was easily loaded. The supplies now reloaded we would take them to the top of the drift and the hand tram them to the face, (our pony, Silver, couldn't get through the low area) some 50m and unload them. This was tiring work, every single piece of steel and wood had to be handled around six times, each time picking up the grey sludge that was everywhere and we sometimes had to lie in it to push the steel arch segments and the 3m split bars (3m pit prop sawn along its length) up that steep hill. Having done that for three loads was enough for anyone, we were wet, sludgy, tired and we deserved the 'ride early' note that Ces Lockwood, our overman, gave us daily for our toils in such conditions, as long as his coal face didn't want for anything.

Another task Paul (Jacko) Jackson and I had was to double back on a Friday night (overtime) and hand load the steel arches the pit would need for the week onto the No.2 shaft cages. This was not as straight forward as it sounds, arches would come in three pieces and loaded on trams in the compound. We had to push those trams 50m to the pit top, not too bad, as the ground was level, through the air-lock steel doors unload them and stand them up on the eight-man cage for someone else to unload in the pit bottom. Whilst the other team below were unloading we would be loading the second cage that was now at surface level and so on, almost a hundred pieces would be handled that night. Norman Allit was the banks-man (in charge as signals and pit top) on these occasions, he could be a fiery and ill-tempered little man, I kept schtum whilst I was near him because I was seeing his daughter Mary. He was very protective of her and would have warned me off seeing her I'm sure. Jacko knew this and kept 'tussing' me and would wind Norman up by asking how his daughter was and brought her into every conversation when he could.

On the Friday night's that we weren't working we would link up with Lenny Hough (my cousin by marriage) and Bernard Fardell and we would spend the night at The County

Arms at Blaby, a well-known nightspot back in the day and frequented by nurses and students from Leicester. Lenny and Bernard were happily married and just enjoyed the night, not drinking much because they were working the next day, but Paul and I had no such ties and restrictions and had fun, our girlfriends thinking we were at work. Life is hard being a teenage boy!

'Born Lucky' Book Two: From Mynah to Miner and Beyond

Getting Serious About Life

Those teenage years were a blur for me as with everyone looking back, Roger and I were closer as brothers, Kate had started work and was making lots of friends of her own, one in particular, Sue Tompkin I remember. They both worked at Ward's shoe factory in the village and always walked together to work, home at lunchtime and home again at the end of the workday.

I had passed my driving test so Roger had Dad's bike to himself virtually, then suddenly he was seventeen and was having driving lessons. I let him get semi-competent and I got him insured for my car, a Hillman Super Minx at the time with a view to letting him drive us to and from work for experience. This worked well and as he grew more confident we would drive further afield, and we would go for a drive at weekends in the morning for a couple as hours, as well as keeping his lessons up. We would always finish the trip out at the Whimsey to see Fred and Lil Hinds (landlord and landlady) at around noon. Fred always had a great sense as humour. If you bought him a drink he would say, raising his glass, "Goo t' hell, I mean good health."

One such occasion, in September 1967, we were homing in on the Whimsey which was on our left and slowing down as the car park was at that time, just past the building, when we both saw Charlie (Chuck) Edwards, a really pleasant, amiable man, coming in our direction and almost at the door. I put my hand up to him and Roger leaned across to acknowledge Charlie's wave but concurrently and unintentionally dragging the steering wheel to the left, the nearside front wheel catching the kerb and the car mounting the pavement. Total confusion! I grabbed the handbrake, but Roger's foot was on the accelerator and the car was heading across the wide footpath towards Charlie, who was now clambering up the three steps into the pub, the car came to a halt before we made contact with the pub wall and as Charlie, coat tails flying, disappeared through the bar doorway. Roger regained his composure, reversed off the pavement and parked the car just as Tony Blackburn launched Radio One playing 'Flowers in the rain' (The Move) on the radio at noon. We sat there and laughed for a few seconds recalling what had just happened, locked the car and went into the Whim. Charlie was on his third pint already! Fred shouting, "What have you done to Chuck, he's a wreck?" Lil stroking Charlie's shaking hand and us laughing like drains as we ordered beers all round, there was only us in! We played darts as usual, we three joined very quickly by Barrie, Norman, the Pete's Rowley and Findley, Barry Gulliver,

Gig Tugby and Jack Walker. A good session was had by all as per usual.

We had many good times at the Whimsey about that time with various landlords but Fred and Lil affording the most laughs, Fred a wiry grey haired comical, lighthearted man and Lil, a more serious often stern lady. One Sunday we were outside as usual on the church bell strike of noon, there being a group of us outside the pub awaiting the sound of the door bolts being pulled back and the Red door opening slowly, as whomever opened it didn't want to be trampled in the rush of customers. Usually it was Fred but on this occasion it was the stern-faced Lil. "Where is he?" we enquired. "He went to a meeting at ten this morning at the Legion (British Legion Club across the road and down a short drive). He'll be here in a minute." We all got our drinks, and our regular crowd of darter's began our game, the pub filled quickly, older guys playing dominoes and the juke box in the lounge could be heard, Lil being rushed off her feet which she didn't mind, but where was Fred? Clearly she had expected him at noon for opening time. Twelve thirty came and in walked Ken Hicklin closely followed by Fred, clearly the worse for wear. "Where have you been?" asked Lil in a less than pleasant voice. "I stayed at the Legion to put my Tote on, my little duck," Fred answered, working hard not to sound too p***ed. "Well, you can get behind here (the bar) and help me now the initial rush is over." "OK my sweet, I'll be with you now." With that, Fred made his way through the now crowded bar to the end of it where the raising hatch was and raised it, this was very close to the darts oche where we were. Fred was unsteady and we were making comments to him, and he was giving as good as he got, always having a quip ready. In high spirits he looked at Lil, her spirits weren't so high, and her stone face showed it. Everyone in the bar was now watching the show. The floor behind the bar had a duckboard flooring for spillage and safety reasons, Fred tripped on this as he said, "I'm here now myyyyyyy preciiiiooooouuusssss!" on tripping, it sent him careering headlong along the length of the bar serving two rooms, bumping into Lil and hitting the far wall radiator with his head, that had above it the 'jug and bottle' off sales hatch window; whose two lady customers had the best view in the house of Fred's antics. Fred was now sporting a multi-coloured black eye and a golf ball size lump on his forehead. Lil left the bar saying, "You can get on with it now!" Lil was not to be seen again that lunchtime. Fred found a sticking plaster from behind the bar, which I presumed he kept for such occasions, placed it crookedly over the small cut on his 'golf ball' and carried out his duties, much more sober than before. He didn't tell Lil that he'd won the Legion tote!

I was still making a good name for myself at the mine and enjoyed different tasks alongside pony driving and supplies.

A face called 24's had been developed in the New Main seam, the supply gate driven to a point on arches taking the coal seam as well as 1.2m of floor stone, the face line driven inseam (no stone roof or floor being taken) and turned left to meet the aforementioned supply gate. This made things very awkward if not impossible to lay rail track for horse drawn rolling stock with that difference in floor levels. Roger had been deployed with me to supply 24's and we had the biggest pony and one of the best behaved, Brownie was his name. Supplies were bought by Tommy Newbold's endless haulage to a point short of the overcast (a place where one road crosses over another) and from there we would take them with Brownie up to the junction with the difference in floor levels. Not good for Brownie as he was faced immediately with a steep incline to get over the road below, on negotiating the incline Roger and I would locker up (secure) the vehicles and rest Brownie even though we had only travelled around 50m. Brownie now had 240m of uphill hard slog to get to the junction, then it was our turn. We had to handball everything up that change in level, around the corner, reload it and hand tram it the rip where Tony Bell and Brian Rawlinson were making the roadway behind the face, Charlie Hannabuss and Owen Halfpenny being in the inseam Stable hole with the AB15 under cutter machine. Geordie Foster driving the heavy-duty trepanner having just finished his 240m cut and bringing that massive coal hungry beast of a machine into Charlie's prepared stable. I was intrigued but told to get away from the face by Jack Abell, the shot-firer, as I wasn't yet face trained and could be in danger. I wanted to be a machine driver from that moment.

Jack Kerr was the Deputy in charge of the district and could see that our system was unsustainable as the face advanced and got further and further away. Jack was a tobacco (bacca) chewing, tubby Geordie and a good practical man. He ordered suitable steel, far bigger than we were used to, we transported it and placed it on that junction. Jack then showed us what he wanted and helped us erect those girders on wooden chocks (supports built from wooden blocks) and so preparing the area for grading the floor for track-laying, in between supplying the face it took us a little over a week. Jack told us that later the manager asked what had to be done to prepare the junction for dinting and he proudly said, "It's done gaffer."

My last tale about my pony driving out of the ordinary days I'm not proud of. It took place in the five-foot seam along an inclined uphill road that had been driven inseam and was called the 'Sister Road' to the Loco Road. It was only five foot high and supported by RSJ's (Steel Joists) on wooden props at either end. Not high enough for any other pony other than Boxer, a wonderful little pony and even his collar used to 'roof' and I would have to pull down on it to get him through. That morning Les Rennocks and I had one load of steel to take some eight hundred metres, but we couldn't do it because our 'drag' had gone missing overnight. A drag is a safety device hung onto the last load draw bar and allowed to drag along the floor to stop any runaways on an incline and derail them at the very least.

After exhaustive searching and much shouting by the in-bye deputy we decided to go without the drag, we'd never had any sort as a problem before. So we set off, me in front checking the track and supports, Les following up behind. That road was slightly uphill for 400m and then got steeper for the rest of its length, on that change of grade there was a bad joint in the track that we often derailed at, a 'natch' as we called them, not this time, straight over. Boxer pulled harder as he felt the load getting heavier on the steeper incline and was using the rail sleepers for footholds, reaching for the next sleeper with every stride, this caused him be bucking slightly but enough to dislodge the limmer cock from the draw bar of the load. Les saw it happen and shouted a warning to me whilst trying to stop the load going back downhill, I dashed back to help him hold the load, it was all we could do to hold that load of steel before we both started to weaken. We were beaten, Les's back was bending, my arms were aching, but we had to get out of the way of the load to let it go free, we timed our exits, and both got out of harm's way. "It'll come off the rails at the bad joint," I hoped, but it didn't. We both shouted as loud as we could "MAINER, MAINER, MAINER" meaning Mainliner......not stopping, it was always used in the case as a runaway underground.

We feared for Albert Sansom and Etna Reason, two old miners working opposite the end of our road digging a water hole, we kept shouting as we chased the runaway load, but it was leaving us behind. We were out of breath and the load was out of sight, but we kept shouting the warning. When we eventually got to the bottom of the road the load was there, off the rails at the rail switch, someone had changed it over after we had left, thankfully. "Are either of you hurt at all?" we asked Albert and Etna. "Not at all lads, we could hear it coming and got out of the way, it was only going slowly when it got here and just dropped off the end of the rails," Albert said, and we found your drag, someone had chucked it in our water hole!" Les and I took that steel in and were thankful that Albert and Etna didn't report us, they could well have been seriously hurt. Never again did I go without a drag!

Darts had become a part of our young lives and it helped us mix with older guys that gave us tips about the game and we became reasonably proficient at the sport, So much so that Pete Findley and I entered the Burchall Pairs, a local well known trophy to play for. We played several rounds and succeeded in getting to finals night at the Union Inn at Measham. We'd not heard of it but talking to the older guys it was 'a big darting pub' and 'some good darter's there, Johnny Walker being the best' they reckoned. All these comments served to unnerve us and come the night we ducked out and did our usual thing on the night. Who knows how we would've done; we'll never know but I was not to do that ever again.

More confidence came and we joined Fred's darts team from the Whimsey and playing in the Barlestone league, quite a strong league at the time. The team at that time captained by Mick 'Gibba' Gibbins and the team being Sid Black, Barry Gulliver, Barrie Bancroft, George 'Gig' Tugby, Pete Findley, our Roger and I. The format being six a side, 1001 double in double out, best of three and the oche set at seven feet six inches. Those Monday nights were amazing, travelling to lots of villages in the area, meeting people and making friends, bumping into old school pals, enjoying the competition, the suppers and the beer. We never won anything, it wasn't important at that time, we were probably a good side to be feared but not consistent enough.

Those times at the billiard hall and the Whimsey were memorable, especially mid-week, Pete, Chris and I were seeing Sue (me), Pat (Pete) and Jill (Chris) all from Ashby-de-la-Zouch and all friends. We would arrange to phone them from a call box at six o'clock in the evening at their local call box. The calls were free because I found that if I tapped out the code and number on the cradle switch, it would connect at no charge, happy days!. Chris, Pete and I would talk to our respective girlfriends in turn, us three never knowing what to say, we must've sounded like real numpties. That chore done we were free to enjoy ourselves for the rest of the night although we'd told the girls, "Erm, nothing happening here, boring really, probably walk home with some chips and watch TV with Mum and Dad." Yeah right, I'm sure they didn't believe us anyway.

The absolute best nights were when it was snowing, and we couldn't get over to see them as arranged. Without fail around six o'clock we'd meet in the Whim, at least six of us and sometimes maybe eight or more, and we'd plan a route around the pubs of Ibstock and set out having a pint or two in each depending on who we'd meet. The route usually being, The Whimsey, The Boot, The Oak (Lovely to see Cousin Pearl and Johnny Wardle), The Crown, The Legion, maybe another one in the Whim on the way past. The Ram, The Ibstock Hotel and finish at The Waggon and Horses, not far to walk home from there. About twice a year it happened, unplanned, cold, bleak at times but thoroughly enjoyable, no one ever said "I'm fed up, I'm going home" on those nights, again, meeting by chance, workmates, school pals, relatives and someone not seen for many a while. It became unmissable. All the landlords and ladies welcomed us on those cold nights, there being not many out and about.

Life was moving on apace now, we had cars, cash and girlfriends, Pete had a Ford Consul, big car it was bored out to 2600cc, bench seats front and rear and 'three on the tree' gears, pale blue as I remember, ideal for three couples, but it did like petrol.

There was one occasion when Pete and I ran out of petrol near the top of Kilwardby Hill leading to Ashby town centre on our way home. It was late at night and luckily Pete had a suitable length as hose and a gallon (4.5l) petrol can. We made our way towards the town centre, finding no garages open we decided to go into South Street behind the Main Street shops. Thereupon seeing a group of small vans in 'Davenport's Beer at Home' yard. We sneakily went into the dark yard and set up a syphon on a van in the middle of the others so as not to be seen from the road. The syphon was taking ages and so we thought it a good move not to be there whilst it was working lest anyone came. Not a good move, we walked onto South Street and had not gone 20m when a voice said from the silence and darkness, "Where are you off to lads?" Turning round and seeing two coppers was the last thing we needed right now! "We're walking home and hoping for a lift," Pete said quickly, I was proud of him at that point. "Where do you live then?" "Ibstock." "You're going the wrong way." "We're going Coalville way, more chance of a lift." The interrogation went on, they weren't satisfied.

"Right then, come on lads empty your pockets on this car." There was a car parked nearby. We had loose change cigarettes, lighters, all normal stuff along with car keys and the vans petrol cap, WHAT! We both thought we'd had it, why had we bought that with us? "Right, put it all back in your pockets and make your way onto the Main Street, you'll not get a lift on back streets, now get off." We couldn't believe our luck, going to the alleyway they told us if we could see a blue light on a building, bloody hell! it was the Ashby police station, we'd picked the worst place in town to look for petrol.

Having got on the well-lit main street we walked a few yards and realised that the car was parked with no lights (they were required in those days), the syphon was still running and we still had to get home. The two of us stood in a dark doorway hatching a plan and executed it perfectly, waiting a few more minutes we found another way onto South Street, well away from the cop shop. Finding Davenports yard still in darkness we retrieved the pipe and can, there was petrol everywhere, must've emptied the van's tank, replaced the petrol cap and made our way dashing from door to door, bush to bush and wall to wall until we arrived back at the car. Just like a Pink Panther movie. With the petrol in the car, Pete set off like the proverbial bat out of hell all the way home and dropped me off at Spring Road. I watched as he was making the turn into the three quarters of a mile long drive to Redholme bungalows when I saw the car stop, I ran to him. "What's up?" I asked. "Out of juice again," he said laughing, "Six miles to the gallon ain't bad is it?" We got the car out of anyone's way and made our way home stinking of petrol.

How good those carefree days were, and they carried on for a while Pete, Chris and I slowly drifted away from our Ibstock group as friends and were strongly into seeing our girlfriends nightly to the point of becoming engaged. Our engagement took place in Ibstock, we had bought the rings in Leicester and discussed where exchanging of rings should be done. Sue was adamant the exchange should be done over a drain and if either ring was dropped it was not meant to be. Who was I to argue? So as we got off the bus at the bottom as Chapel Street, it was dark being the first of October, and made straight for the telephone box at Mee's bakery, there happened to be a road drain near to it and we made use of it and to our satisfaction nothing amiss happened during the exchange of rings. We went into that telephone box for a snog to celebrate our commitment to each other, 1966 was a memorable year.

Things became more formalised after that, we would dress up on weekends, us lads in three-piece suits, winkle picker shoes, ties and fob watches, cuff links and tie pins. The girls with bouffant hair styles, perfectly done, Mary Quant dresses or mini-Skirts, high heels and stylish make-up with jewellery to suit. We thought that we were the height of fashion as we entered pubs and clubs at the weekends, us three lads sharing the driving before the breathalyser came into force. Life was still enjoyable but different, not so wild and spontaneous, but more planned and thought through, the girls would have the main say as to plan for the upcoming evening, but we lads insisted on finishing on Sunday night at the Whim, singing in the car as we headed for home "You go down through Barlestone along to Nailstone, Ibstock City looks oh so pretty. Get your kicks on Route 66". We thought it sounded great after a few beers, but the girls weren't so keen. Why is it when boys take out girls and are enjoying themselves and laughing, that the girls are often not so keen, having consumed equivalent alcohol, and mention your performance when recalling the event later? It gets worse with time I'm sure and the ladies never ever forget an incident, how do they do that?

This carried on for a couple of years or so and was a brilliant time but the ladies were maturing faster than we guys and wanted to start saving and planning for our weddings. Us lads were quite content for this to carry on, as most are, but now things were getting slightly more serious and all three of us were persuaded by various underhand means that putting cash aside for such events was the way to go.

I was the first to get married in July 1969. It was amazing, like an out of body experience, I didn't really believe it was happening to me. All those friends and relatives, everyone there celebrating and wishing the two of us well for the future, and maybe thinking we were too young, Sue was nineteen and I was just turned twenty-one.

Wedding day

Maturing Quickly

For the next three years, life was more difficult in that Sue and I were working hard to put money aside for a deposit on a house and also to furnish it, but helped greatly by the fact we were living with her parents Arthur and Joan. They were very good to us with advice and practical help, I contributed to the household with fuel, decorating and gardening, neither of which Arthur was keen on. Arthur and I became drinking buddies and Joan enjoyed a libation or two and with Sue, we'd often be out and about at weekends as a foursome, gradually drifting away from Pete and Chris along with their partners unintentionally.

I was getting on well at the mine, keen, organised, skilful and an achiever was the reputation I wanted to foster, and Pat Martin was helping me to achieve it, he already having that reputation. What a stroke of genius whomever thought to put us together, I'm sure it often happens that workmates become best friends this was no exception. As a pair we tackled numerous awkward and dangerous jobs together and as part of a team, but we would always take charge.

Our Dint header was going extremely well and setting three-metre-long beams on two-point four metre legs at one metre spacings and we were fast approaching a forty five degree junction turning to our left. The four carriers (big steel rsj's set in pairs) were on site for a four-way junction. "But we're only making a three-way junction." I said to Maurice Twigg, our regular deputy. He checked with the surveyors, and they agreed with me but the steel had been sent in for the original junction, which was a four way. "How're we going to do this now". Said Mo slightly panicking. "Leave it to us". replied Pat and I along with Jimmy Mee our third man. "You go outbye (away from the workings) and sort tackle with the supplies team (Albert, Fred and Jack, if you remember).

With Mo out of the way we built two chocks (like Jenga stacks) at each side of the gate at an estimated forty-five degrees to our centre line and placed a pair of the big rsj's on top of them. Jumping on the machine I started cutting in at right angles to the rsj's (Carriers) the dint header devouring the coal in front of it and spewing it out onto the conveyor behind. Junction cross girders are longer than normal roadway supports, and these were four metres long and so I had to cut that far past the carriers. Maximum permitted distance being two metres on the managers support rules. So we had to hurry before anyone caught us. I had cut the width of the machine, two metres wide, and four metres deep plus a bit wider at the base for the splay of the legs. Pat, Jim and I carried in a cross girder and threaded it over the carriers, Pat and Jim setting the leg to it whilst I held the girder above my head and in place. Now for the second one, no problem, the two settings were strutted together, timbered over and secured. We were naughty boys doing all that under unsupported ground, it was a calculated risk, and we knew it, but we weren't caught, there were no accidents and Mo was pleased with the result. No one asked questions on the surface, they were just interested in progress. That was an easy junction after that, it was done in three shifts.

Whilst living with Arthur and Joan we were saving as much as we could for a deposit on a house. I had ideas of emigrating to Australia or Canada, but Sue wasn't having any of it. "If I could I'd have a house built at the bottom of Mum's garden," she would say. I wasn't going to win that one, so I didn't persist. We visited a building site nearby and saw plans of houses being built in an area known then as Beaumont Gardens and paid a holding fee giving us preference to buy. We were so excited and told our parents, both sets were pleased but very apprehensive about the mortgage risk facing us. The house was built on Huntingdon Road and after being 'gazumped' to the tune of six hundred pounds on a four thousand pounds house we moved in in February of nineteen seventy-two. We'd worked so hard it was fully furnished and carpeted throughout, what satisfaction!.

For the previous two years I had been at Nottinghamshire Polytechnic College and was really into the education side of mining and have many fond memories of that time. The travelling back and for was a daily adventure, the long lunch breaks (one- and three-quarter hours) where we would go into 'slab square' as locals called the area in front of the council town hall and listen to people spieling about all sorts of subjects. We would visit the local courthouse and observe justice in action at the assizes from the public gallery. Those lunch breaks were far too long and we negotiated with the lecturers to shorten them to an hour so that we could get away earlier and that was mutually agreed. I enjoyed all the subjects at NottsPoly (now Trent University) but surveying stood out for me. Based in a room near the Arboretum Mr Robb would hold his lectures and then take us into the Arboretum for practical use of the Miners Dial and theodolites. It was so comical, on entering the Arboretum, there was a small pond on the left, half surrounded on the road side by trees and bushes. Ducks would be seen from time to time on the water doing their thing, but other times it would be clear of ducks but strangely quacking could be heard. On investigation we were told of an escaped Mynah bird that had made the pond it's home and had learned to quack like a duck, it was so funny to hear it after we knew the tale and always looking for it as we passed.

We were settled in our new home, and things were going swimmingly, I had built a garage, completed the driveway and gardens front and rear were presentable and our neighbours were brilliant all around us. Things had been a little tight financially but not unmanageable, as I was back at the mine and making overtime and Sue was earning good money as an overlocker at a local hosiery factory. We started socialising more, firstly with the neighbours, mainly Saturday nights at each other's houses and were used as 'get to know you' exercises that carried on for years because we got on so well together. Sue went out occasionally with our neighbours for meals and bingo outings as well as shows, whilst I would go out playing darts for a local team from the Plough on the Green, Norah and Ted kept it in those far off days.

Those darts nights were great, we were in the Ashby league but never winning it as the War Memorial Club were much too good, although we gave them a run for their money. We did win once, the coveted Hospital Cup, a really ornately beautiful trophy. However the Ashby league collapsed, and we applied to join the Woodville Darts League, a much stronger and better organised league of three divisions. We were placed in the first division by the secretary George Bannister that ran the league very efficiently amid protests from other teams that thought that we should start at the bottom of the league, I think I would have said the same.

Joining that league meant more travelling but fortunately Charlie Tissington had bought his football team to the pub, now run by Tim and April, and said that we could have a loan of his team mini-bus on Monday nights. On darts night we didn't just play darts but dominoes too, players could play both games but we had enough players to man (and ladies) up both teams plus wives, whom did the suppers in turn at home games. This meant the mini-bus was extremely full, Dave (Black-dog) Earp and I had arranged that I would drive there and he would drive back.

Loading of the bus was the usual kerfuffle of people wanting to sit next to someone else on the bench style seats that ran the length of the bus with no armrests and some older people struggling to get in and needing a helping hand. All loaded and accounted for and back doors locked, with a struggle I might add, and so I set off on our first away match journey. On arriving at the first set of traffic lights I braked gently, as I thought, the brakes locked on and the bus came to an abrupt halt, my god those brakes were good and it solved the problem of the rear doors being under pressure as the passengers all slid towards the front of the bus, bunched up.

It became a standing joke each time we took that bus out and put both teams in good spirits for the forthcoming games. We won that league four years on the trot from the off with players Like, Richard Rennocks, Tommy Logan, Danny Povey (Dad), Phil Lilley, Charlie Pitchers, Ian Warhurst, Pete Marsden, Brian Abbott, Pat Rastall, Johnny Webster, Arnie

Forrester, Dave Earp and I (captain). It really was a pleasure and honour to captain such a team with some teams bearing a slight grudge, it would all the more rewarding walking away with a win.

Whilst playing for The Plough I teamed up with Dave Earp in the pair's competition, we got through some very tight games and one in particular that we should never have won at The Rising Sun, Woodville, the losing pair not happy when I hit the winning double. That took us to finals night at the Hartshorne Working Men's club, here was the night where darts singles, pairs, triples, divisional cups and dominoes cups were played for, a very busy night. Dave and I got through the semi-final round and were to face a team from The Turks Head, Donisthorpe, David Nish's pub of Leicester City and England. What a match that was!.

Best of five, double in and double out, we were the first final to play and with people gathering we started as the chalker shouted 'Order for darts, game on'. It was tense and the first final I'd ever been in on such a night, and it went to two games each and they had the darts in the final leg. They failed to get off and I hit a 152 (double sixteen and two treble twenties) then they got off and were chasing us down the 501 points required to win. Then they started playing unsporting games, not illegal but unsportsmanlike, they would walk straight back to our player on the oche rather than to one side or walk slowly to retrieve their darts, taking them from the board singly, slow down play, speed up play and anything to disrupt our rhythm. Dave and I dealt with it.

I left him double sixteen to finish, he made a right pig's ear of it and left me double seven. Waiting patiently six feet back from the oche for the dawdling opponent in front of me to get away from the board and behind me, I used that small amount of time to compose myself, stepping up to the oche I visualised the dart going into double seven, threw it and hit it, I threw the remaining darts to the floor, turned to shake hands with the losing pair but they had turned away, shame on them. Nishy came into the club as Dave and I were about to refill our glasses, "When are you on Don?" he asked. "We've been on and beaten your pair in the final Dave, good match though". I said "Oh dear, I bet they're not best pleased, they don't like losing. Here, let me get those". He paid for our beer and circulated amongst the crowd. The beaten pair found us later, apologised and shook hands, too late, I still had a bitter taste in my mouth.

Meanwhile back at the ranch, having done my face training, I was now a machine driver, cutting a two-hundred-metre-long coal face in good condition and it was in the New Main seam, but only a metre high and a lot less than that in the chock track taking the chocks beam, base and debris into account. A good average would be two traverses of the face, often less but sometimes more, there were a lot of unforeseen variables to winning coal. Problems could be methane gas, bad roof, floor lift, fast panzer, machine fast under low chocks, weight on the face and a myriad of outbye (away from the working face) problems, nothing would be unexpected.

When everything was running smoothly the belt would be running, the pre-start alarm would sound from the stage loader and panzer, I would turn on the water to my Double Ended Conveyor Mounted Trepanner and a few seconds later, my gloved hand would press the green button and all five of its cutting elements would spring into life, pushing in of the 'haul' lever would set the coal hungry beast on its journey to the two hundred metres away main gate. I preferred cutting towards the main gate as the coal flowed away from the machine whereas cutting the other way, towards the supply gate coal had to flow under the machine and lumps would endanger exposed hydraulic jack hoses and possibly damage them, another source of delay.

Everything was running as it should do, I would be watching the mighty machine pushing wooden props off the panzer that had been set to hold it down and me taking them off the conveyor for re-use and placing them in the crawling track of the chocks whilst steering the machine; constantly shining my cap lamp at the roof turret and the fins of the trepanner bucket keeping it within the seam and leaving enough coal tops up to strengthen the weak mudstone above. This meant just skimming the soft fireclay floor at the base of the seam, cutting any amount of that fireclay would mean that the product I was sending to the surface would be inferior and may be rejected by the Electricity Generating Board. Lumps of stone were really no problem as they can be screened out more easily than fine dirt such as that cut by a machine. That dirt has to be washed out and does affect the product adversely and increases the ash content which the electricity boards don't like outside their set parameters.

Trepanner coal cutting machine

A shearer coal cutting machine

Shearer in action

Coal face chocker

Coal face machine

Getting on with my task I could see behind me Johnny O'Grady snaking the panzer over into the newly cut track, I can see him leaning over the panzer, shining his powerful snaker's cap lamp along the new panzer line. Johnny was an excitable Irishman that talked at such a pace he couldn't be understood at times, but what a workman and with such standards. Johnny was snaking props on as the panzer was rising slightly. I could see everything was going fine, the noise was as usual, the haulage chain banging on the panzer flight bars, people shouting on the tannoy, chocks scraping along the roof whilst being advanced, Johnny shouting at the chockers, Melvyn Green and Johnny Ramsell, that they were pulling the panzer back and spoiling his panzer line. I could see their lights too alternately looking at the base of the chock and it's canopy, ensuring it was nice and flat up to the newly exposed roof when it was fully advanced and set. This environment was satisfying to me, a full panzer of coal, the beautiful sound of the machine cutting and whining from the haulage end akin to a jet engine plus the constant hiss of the water sprays. I enjoyed that part of mining, physically hard at times but satisfying.

One thing I didn't enjoy along with everyone else was when I'd get to the end of the face with my machine was the sudden loss of power. "No power to the machine". I'd shout on the tannoy. "Check the panel in the supply gate." "It's showing a red light on it Don, I'll try and reset it, I've got a key." Ken Beasley let me know. Ken was a bull of a man, well built and very strong, you'd want him on your side in a scrap, but one hell of a willing worker. "It's no good it won't reset, I'll let Trevor Jordan (Electrician) know, he's down the gate road". Ken reported. "Melvyn, nip back to the halfway mark and check the cable for me". I asked of the chocker. The halfway point of the face is a notorious place for damaging machine cables, it's where they flex, bend and turn over to allow the machine to pass on its next cut. "It's skinned for about a metre Don, we need a new cable."

"We'll let Trevor decide that," was my answer to Melvyn, knowing what Trevor would say. At least the machine is in the most convenient place for a cable change, the cable and its handler being stretched out rather than overlapped.

"Ken Beasley, get me lots of cover boards ready and when I let the panzer go put one on each flight, you'll need a hundred, get the rippers and the bidi men to help you." Ken does as I asked, and I let the panzer go, knowing that when I stopped it there would be boards from the machine to halfway at 1m spacings. These were to raise the metal hinged cable protector clear of the pan-side trough and support it, a difficult job crouched over it whilst on your knees in restricted height. Next is to knock out the roll pins that retain the cable. "Come in Don, It's Trev, the cables down (Damaged) I've tested it. We've got a spare in the gate, I'll get the rippers, bidi men and Ken to help me get it to the face line." "OK. Trev, we'll get the old cable out of the cable handler with the fitters and the machine team" "Come in Winnie (Ray Winfield) bring your hammers and punches (Dog head nails) on to the face from main gate to halfway. "Been listening Don, me and Brian (Henshaw) are on it," Winnie answered. "Come in Don," shouts Jackie Bateman. "The gaffer's on the phone going mad, wants to know how long you'll be." "A f**king lot longer if he keeps stopping us to answer silly questions," I snap back. "I'll let him know when I've a clearer idea Jack," I added more calmly.

An hour has passed, and the damaged cable is freed and placed in front of the chocks for retrieving on the night shift. That's the easy bit done. "Trev is the new cable ready?" I asked. These trailing cables are 260m long and run the whole length of the face and into the supply gate. They are around 75cm diameter with a monstrous pommel (plug) at either end plus they are extremely heavy and awkward to handle, not unlike wrestling an anaconda at times. "Just about, give us ten minutes, we're dragging enough onto the panzer so that the cable weight will pull the rest of the figure of eight cable on to the face." I liked his thinking, much safer than tying the pommel to a flight bar and much less chance of damaging the new cable. Although there was still that possibility. "Just hotch the panzer for a start Trev until it's (The new cable) going OK." "Don't try and teach your grandma how to suck eggs. It's not my first day out of the siding you know," Trev retorted. Great, I loved those sorts of confident answers.

Trev sorted the new cable and tested it before it was plugged in, those pommel winder mechanisms were definitely not ergonomically designed. The roll pins put back and the metal 'Bretby' cable handler dropped back into the pan-side trough, the machine tried for power and we're ready to go again. "Let the panzer go Jackie and let his lordship know we're coaling again." That took a total of three hours of hard work and trapped fingers to get back to where we started from, I hated it!. The night shift won't be happy either, they'd have to free the old cable along the length of the face, run it off and load it up for transportation, an awful, frustrating and tiring job for no reward.

Life was good at home, finances under control, great social life, nice car and job satisfaction. Sue broached the subject of children, something I had thought about but not deeply, and we decided it was the way to go. We were very lucky and in no time at all Sue was pregnant. I was attending every ante-natal class with her and as she got near her time it was at one such morning class that Dr Happel showed some concern about her blood pressure, phoned for an ambulance that whisked her off to Derby City Hospital with me in hot pursuit. That was on a Thursday, Sue's blood pressure didn't subside overnight and so the doctor thought the risk to the baby was too great and he decided to induce her. Later that day at ten past six to be precise we had a beautiful baby girl, it was Friday the sixteenth of April nineteen seventy-six and Good Friday, It was an emotional time for me, why are men reduced to tears when their child is born?. I went to the waiting room after the birth and 'Crossroads' was on the TV, but I couldn't see it clearly because I was blubbing intermittently, thank goodness I was alone.

My how that changes your perspective on life!. I was acting under manager around that time, a high-pressure job for me as I was officially a deputy but acting overman and now acting under manager through unavoidable circumstances at the mine. I would come home and see Andrea lying on the sofa, propped up by a cushion, gurgling and waving her arms and legs whilst having the biggest smile and I knew why I did what I did at work.

Sue and I went out most weekend nights, very willing grandparents doing the babysitting honours, mainly to the local pubs in Ashby. She continued with her keep fit meetings and social gatherings and I went out once a week with Pat and Maurice on a Tuesday. I would pick up Pat and then Maurice and we would go to the George and Dragon at Thringstone when Betty Henderson kept it, she being more of a Blackpool B and B landlady rather than a pub landlady, but nevertheless well thought of.

On one occasion we also took Roy with us, Pat's brother-in-law, Pat had warned us that he was a 'lightweight' drinker. All went well, we were playing dominoes with Jack Rush and driving him mad putting fours next to fives and generally messing around and having a good time, laughing loudly. Roy was watching and laughing with us and taking a turn at dominoes. Three pints and Roy was talking to all the regulars, looking at different fixtures on the walls and shelves and remarking on them, not in a complimentary way and watched by Betty. As he made his way around the room he tripped on a stool and knocked into a table spilling drinks, he was well on his way. Roy was a member of a brass band at the time and as he came from the toilet he saw a post horn hung high up on a roof beam. He stood on chair and took it down planning to play it, slipping off the chair and narrowly missing another table he managed to stop himself falling, recovered his composure and managed to play a few notes of The Post Horn Gallup. It was too much for Betty, she shouted him to stop and ordered our group out of her pub for being too rowdy and causing a disturbance to her regulars. Roy came out with us a few times after that, but we kept an eye on him.

When Pat and I were on Friday afternoon shift that finished at eight forty-five we would visit different pubs for a few pints with workmates, these were generally high spirited affairs at local pubs such as The Castle, Birch Tree, Ellistown Hotel and

The Halfway House but one such place was The Liberal Club as it was then known. A small club placed at Marlborough Square in the town of Coalville it had once been a lovely plush place but had declined, but it did have two snooker tables in an upstairs room.

We would challenge each other throughout our shift and look forward to our game, shooting through the showers at warp speed, as everyone did on Friday afternoon, mascara evident (black rings of coal dust around the eyes, missed in the rush) we'd head for the club. At the bar Pat, being a member, would order the beer, cues and chalk and off we'd go up the stairs having drank half of the pint we'd bought. Once in the snooker room and the table lights on we'd drain our glasses and place them on the dumb waiter with some cash, lower them down and land them at the bar raising and lowering the conveyance making the glasses rattle to get Charlie's (barman) attention. Then we'd set the balls up on the very worn table, that done, and two more pints returned the game would commence.

Playing and laughing about the shift just elapsed, the beer would be most enjoyable, and we were soon at the dumb waiter again, rattling the glasses for Charlie's attention. Continuing the game, the shots becoming more adventurous and outrageous, more beer available on the conveyance and everything was well with the world. Glasses on the dumb waiter again, "Same again Charlie," Pat shouted down the small shaft. "Ugh, OK," answered Charlie with more than a little angst in his voice, ignoring that, we got our beer and carried on with the game.

The tables were very poor, the cloth very worn and the cushions were 'dead' so much so that to 'double' a ball across the table required maximum cue effort, to do something similar along the length of the table would require a 'run up' but that didn't bother us, we were enjoying ourselves. Another beer was the order of the day, rattling the glasses for the fifth time in short order, we hadn't finished a frame yet. Pat didn't have chance to shout down, an under pressure Charlie was out of his collar and shouted up "I've got more customers that you two up there, you can bloody well wait your turn." In the background could be heard an urgent "SHUSSSSHHHH" from the bingo players. He made us wait for that pint but we laughed about our vision of a 'rattled' Charlie.

'Born Lucky' Book Two: From Mynah to Miner and Beyond

Moving Up

Everything was good for us as a family and was working well, Sue felt that she could cope with another baby and again we were very fortunate not to have to wait at all. I was working afternoon shift at Ellistown on Friday ninth of June nineteen seventy eight on sixty seven's face when a call came from the control room over the tannoy "Don, come in Don" it was Alvin Andrews. "Yes Alvin, what do you want" "Your Sue's been on the phone, and she says she's in labour and can you come home straight away." "On my way Alvin, ring her back and let her know please mate." "Will do Don." With that I crawled off the face to a big cheer from the face team. As I walked outbye to the pit bottom, going through my head was ' don't rush, be careful driving, no good having a crash on the way home'. Excited as I was I kept it together and got home around six thirty pm, Sue was there with Pauline Tebbs (next door neighbour) and Judith Fothergill her best friend, both comforting her and keeping her calm whilst caring for Andrea.

I whisked Sue to the local cottage hospital with her case of paraphernalia and was greeted by nurse that took us through to the maternity unit. The unit was strangely quiet, I had expected sounds of babies crying, doctors/nurses talking and women screaming at their husbands for putting them through this ordeal. Sue seemed to take it all in her stride but suddenly remembered something she'd left behind and asked me to nip home for it. Being less than a mile it was no problem, and I was there and back by ten past seven, fractionally too late, Sue had just delivered our baby son Stuart. What luck, one of each, we were over the moon at our good fortune. Born lucky or what?. Sue was now exhausted and needed rest. I was buzzing and with Judith taking Andrea for the night I headed straight for the Plough. No paternity leave in those days and with lots of help from willing grandparents I muddled through until Sue came home a week later.

Sue with Stuart and Andrea

Sue and I

Things were also going well at the mine, I was now an overman and taking life very seriously at home and at work. I had this urge every shift at work that I had to beat the previous shift at everything and if that target wasn't high enough I had to beat the best performance for that operation.

I have said previously (Born Lucky but my life was the pits) that I enjoyed working the mine holidays and I did. This shutdown was for a week, and I had volunteered to work the afternoon shift. Fred Johnson was the under manager in charge for that week and he was on day shift as normal. The task in hand was a face-to-face transfer of equipment and the operation was in the support transfer at the time. This meant at the worked out face the guys would be freeing the hydraulic supports (chocks) setting wooden permanent chocks as their replacement. Then transporting them through the face line under the cover of the yet to be freed chocks and load them ready to transport to the new face by another team of supply lads. These supply lads had to take the loaded chocks via three separate haulage systems to the new face line for installation by another team of men.

I was the section ten man as always (In charge of the whole mine in the absence of senior management) but signing statutorily as a deputy for a district, the installation district in this case. Looking back through previous reports I saw that three or four chocks per shift was being achieved. Lawrence 'Loggy' Hudson was the salvage district deputy and as we all met in the deputies cabin in the pit bottom I told them, "Your target is five a shift, five off and five installed and I'll see what time it is when we've completed that, if it's a reasonable time I'll let you go early for a pint and pay you a full shift." Arthur Houghton, the charge man on the withdrawal team, was all for it as was Jack Tonks the installation charge man. They'd give it a go as did the supply lads. Men motivated, now all it needed was organisation and timing. The supply lads were key, Kev Ensor, Teddy Lines and Ian 'Acker' Adkin were the best at achieving targets and I was in constant touch with them. Loggy was brilliant as ever (he was my vice-captain in the rescue team) and we achieved the target with forty five minutes to spare. I let the guys go early, I was happy, they were happy, and Fred was happy when I phoned him with my end of shift report, ending with "I'll go and get a bath and a cup of tea now then Mr Johnson" although I was already bathed and drinking tea in the control room, just to make him think that I hadn't come out of the pit too early.

All went well until Wednesday afternoon when Brian 'Bungey' O'Meara had informed Loggy, his deputy, that Ron Tugby on the day shift had complained that we were making his shift look fools and that he didn't want to do five again today. On being told this my reply was, "I'm here for a full shift, you can be too it doesn't matter to me." All was well again, and we got our five for the rest of the week. I had long since learned that a pitman will work harder for a short shift target than ever he will for money, on a short-term basis. It was to be a big shock for me in the future!.

Another time that I used the 'Job and Knock' incentive was on a system change at a supply gate rip. The object being to strip out nine chocks and a Webster Packer (a piece of machinery for moving ripping dirt sideways to fill a pack hole behind the chocks for roadway stability) and replace them with nine new ones with cam packer boxes attached to their rear beam.

I was on afternoon shift again and again it was holiday time. Fred Johnson was once more in charge, over the previous week the old chocks and the Webster packer had been stripped out. But all through the weekend not one new chock had been installed. Mr Johnson was making me aware of the situation, still no chocks were installed by Monday day shift. The first chock was on site with another in the pit bottom, "Where's the next one Mr Johnson?" I asked. "You won't need it Don." "Where is it?" I asked again. "You won't need it, you won't get that far." "I won't ask again," I said feeling my neck getting hot and Fred could see it. "It's on the pit top, ready to go in," he replied. "Get it in the pit for me please before the surface men finish their shift." "Will do," he said phoning the pit top to make the arrangements.

I knew the small team I had to work with and as always they were good men. I wanted three chocks on that face line that shift I told them after seeing what was reasonable plus a bit more with the promise of an early finish. Aubrey Rudin and Mick Gibbins (Fitter) were set out to instal the first chock. Alec Baillie and Jimmy Lawrie, two excellent Scots, were transporting the chocks to the face using an endless rope haulage with a young Stuart Baillie driving the haulage engine. Johnny 'Tutty' Redfearn was the loco driver bringing the chocks to the district with Alan 'Giddy' Southam as his guard. I surmised that previous shifts had either lack of good supervision or had been 'lying on it' meaning swinging the lead, I wanted to embarrass them.

The first shift went well until around seven o'clock and with two chocks on the face Alec Baillie rang me at the face, they had been waiting for Tutty and Giddy and had gone outbye to meet them only to find the chock stuck fast and the Becorit (Floor mounted monorail) track badly damaged. Having no new track available we were well and truly stumped, all we could do was back the loco off and remove the length of damaged track in preparation for the day shift to replace it.

On the next shift, Johnny Edwards met me in the surface deputies cabin with the report I had written the afternoon before. I had written the story of the shift and finished with 'Nearly had three on the face but TBF'd'. "We couldn't work out what TBF meant but when Arthur Tyler saw the loco he phoned me and said it meant Track F**ked and Bosted," Johnny said laughing. We had a good laugh about it and I explained the accepted meaning of TBF. Overmen and deputies are not allowed to laugh unless everything was going well, which it rarely was. My lads lead the way again all week, I rewarded them the only way I could and they rewarded me by enhancing my reputation as an achiever. I liked to think that all the men under my control over the years did what they did for me, not just because they were paid to do it.

Time marches on, Andrea and Stuart were now at Hilltop Primary school, Andrea being selected to represent the school along with one other at Buckingham Palace on the Queens jubilee celebrations, being presented with a mug and commemorative coin. Sue was up at the school at every opportunity and I helped at fund raising events and at Christmas time where Stuart played one of the three wise men. I always plumped for the 'roll a penny' (well 2p actually) stall, having a box of sweets by me, I would give away sweets to the kids whether the penny finished within the squares or not. Word soon gets around and they would four or five deep around my table, I loved it especially when Mum's bought their not yet school age children to play, the joy on their faces stays with you forever.

Nineteen eighty-four and there was a strike in the coalfields. At work the shower water was stone cold, with there being no boiler men at work, and there would be big cold condensation droplets on the high ceilings ready to drop on an unsuspecting overman or deputy passing naked underneath. It's unexplainable what that feels like on warm naked flesh as we passed between clean and dirty lockers. We were expected to work for safety reasons through the disputes but still needed a shower after our shift. It was awful showering in cold water mid-winter, and I decided, like others, that I would go home to shower. I was in the shower when Sue bought the kids from school, sending Stuart to the bathroom saying, "Go and see your Dad Stu, see how black he is." Stuart came in to me and just stared in disbelief, didn't say a word just couldn't take his eyes off my black face. I'll never forget that look, it was as if to say, "If I got that dirty I'd be for it." I decided to brave it out in a cold shower, it wasn't right putting clean clothes on in the baths on top of pit muck, but those condensation drops would have half-filled a tea cup!

I was now senior overman at the mine, as high as I could go without going into management. I was offered the position of afternoon shift under manager, but I couldn't see myself on afternoons regular for the remainder of my career, some thirty years at that time or so I thought.

Pat and I were still causing havoc on our nights out. I remember an inter-pit competition of darts, dominoes and cribbage being arrange at Snibston Miners Welfare Club, these were always fun nights. Pat and I had entered the darts team of six players, 501 double in and double out, best of three. The game went to the deciding leg, it was nip and tuck down the final double and we had ripped ours down to double one with me throwing next. The board was surrounded by onlookers supporting both sides, I threw my first dart over the top of the double one, the next went into the single one but was hanging loosely near to the double, instantly I threw my third dart, knocking out the loose dart and hitting double one. There was uproar, our supporters cheering for our win and chanting "Who are ya" and their supporters crying foul. We knew we had not won on that throw but kept the rumpus going for the sport of it for a while, stoking the fire, as they say. They finished the game hitting double five on their next throw and then it was our turn for the flack.

I must tell you about another occasion involving a pool game. My very good friend Peter Marsden by chance asked me if I'd partner him in a pair's competition at a venue and date to be arranged. I wasn't a pool player regularly but enjoyed a game of snooker, even so I agreed to partner him. Pete was running a training course in London somewhere for the CEGB and was staying over for the week when the date and venue became available. I said to him," Cancel it Pete, I'm not too bothered." He said, "No, I'll drive home on Thursday night, play the match and get up early Friday morning and drive to London in time for the course."

That was settled and the night came, eight o'clock start at The Bear in Swadlincote, best of three, winners move on to the next round on the same night. We were drawn in an eliminator round and were first on the table, balls racked up, referee called play and I broke off, the game went to and fro but they took the spoils. Second frame went pretty much the same, but they were left with only the eight ball on the table and with a reasonable pot, they missed it. It was my turn; they had left almost a straight pot to the middle pocket but at a slight angle to it. I had long since learned the lesson not to rush into these situations when an easy victory is in sight, so I surveyed the table and the possibilities whilst casually chalking my cue; getting down over the shot I could clearly see the line I wanted and with a little stun on the cue ball I played the shot. Oh bother!. My cueing hand hit the table edge just as the cue tip hit the cue ball, the cue ball went forward, made contact with the eight ball and sent it towards the pocket, but it never made it. They were left with a similar pot to mine but much closer, a mere formality for them.

Every time that pool was mentioned over the next forty or so years that bloody big Yorkshireman from Brighouse never failed to bring into the conversation that game; and how he drove two hundred miles round trip for a fifteen minute pool game. How I wish we could do it again with him.

In 1985 our family had a devastating blow, my sister Kate passed away with cancer aged 34 leaving her husband Barry and her three girls aged three, eight and ten. The local council were very good, letting mum and dad move in with granddad and Barry and the kids have mum and dad's house. It meant that they had adjoining gardens back to back, and that mum and dad could watch over the girls whilst Barry went to work. Then another heart-breaking tragedy, five years later Barry passed away in the doctors surgery of a heart attack and again leaving the girls, Rachel now fifteen, Ruth thirteen and Ria eight. The council again very sympathetic, with mum and dad applying for guardianship, they allowed enough time to pass until Rachel reached sixteen at which time they issued her with a rent book for the house, this saving the girls from going into care. Rachel grew up overnight and shouldered the responsibility for the house, her siblings and their upkeep. With mum and dad watching over them they made a success of living together and are now living well balanced lives their mother and father would be proud of.

'Born Lucky' Book Two: From Mynah to Miner and Beyond

Kate

Rachel, Barry and Ria

Barry, Kate, Rachel, Ruth and Ria (baby)

'Born Lucky' Book Two: From Mynah to Miner and Beyond

Good Friends and Times

I sometimes think we can measure personal satisfaction by the friends we keep through the years and what we do for each other and our common interests, whether it be work, sport, games or any pass-times. In my life the major interests I've had include my work, family, motor cycling and racing, cars and racing, darts, squash, keep fit, social mixing and of course golf. These are all activities that involve different groups of people and enrich your life with their presence in it, the longer the better. All the more satisfying when you lose contact with them for a while and then re-connect and take up where you left off as if no time has passed.

As I have mentioned in my previous book (Born Lucky but my life was The Pits.) My more serious spell of motorcycling came to me in my very early thirties when Sue had passed her car test and I was given a bike by a friend for some decorating I had done for him. The 125cc Honda served its purpose but I needed a 250cc and I bought another Honda, a 'Dream' this time. This meant Brian (Abbott), Pete (Marsden) and I could venture further afield to various venues of our choice, but not too far in the time permitted by our good ladies, Jacquie, Tilly and Sue respectively. We had boundaries you know.

I guess it really started in nineteen eighty-two, by that I mean talk between the four of us, Brian Abbott, Peter Marsden, Tom Fothergill and I about a trip for a few days to the Isle of Man TT races. It was to be arranged by Brian, our co-ordinator and organiser. It was with great difficulty that he'd sorted ferries and accommodation, we being new to the situation hadn't realised just how popular this event was. It finally was arranged for a four-day trip, Tuesday to Friday in Brian's Vauxhall Cavalier, travelling to Heysham and across the Irish sea to Douglass.

It was a great start to the trip, finding our digs near The Terminus in Douglass, getting advice from Rob the hotel owner as to where to watch and how to get there. He then told us that Mondays race had been moved to today because of bad weather and that we could catch the second race in Douglass at the bottom of Bray Hill nearby if we looked slippy. Arriving at that point, it was a five lane ends layout and the only thing preventing us getting up to the track was a marshal and what I can only describe as show jumping tressels. We ran downhill the few yards to the crowded tressel area to get a glimpse of the action, something whooshed past but we were not yet close enough to see. The road was quiet as we made our way to a viewing spot amongst the crowd, I could see to my right up Bray hill and the sound of a bike, flat out approaching from the Glencrutchery Road, over the lights at St Ninians and then

No.11 came into view, it was Mick Grant, he got level with me and I had never seen anything so fast in my life. In a split second he was gone but not before he had clipped a manhole cover at the bottom of the dip that sent his machine into a speed wobble, a real tank slapper that almost lasted to Ago's Leap (Giacomo Agostini) before his front wheel was pawing the air at that spot and disappearing over the hill, making his way to Quarter Bridge, never shutting the throttle. I turned away, shocked at what I'd just seen, I was amazed, numb, dumbstruck, confused but also utterly hooked for life at what I'd witnessed. These guys weren't right, there was something missing in their make-up surely!. Whenever I recount that experience the hairs on the back of my neck stand up as they are now whilst I'm typing, such was its impact on me.

That first impression was in fine warm weather, tomorrow was forecast similar, so we made our way back to The Edelweiss Hotel for a meal and to formulate tomorrow's plan and also having time for a few pints of Okells ales before our first race day proper. The decision had been made to park inside the track at Braddan Bridge so we could view from there and move to Union Mills for another view of the action during the race, all the time listening to the race commentary on Manx radio whilst travelling. I should say that there was no way to cross the track from inside to out or vice versa, it's forbidden from race start to the end of racing for the day. The plan totally fell apart, it being poorly executed, the morning race was delayed due to mist on the Snaefell mountain, not because the riders wouldn't race in it, but because if a rider should come off his machine, the helicopter may not be able to find him/her. Yes women do race around that thirty seven and three quarter mile circuit!.

Isolated at Braddan Bridge and the roads closed for racing we decided to walk on the back roads to Union Mills to the church hall we were told about by Rob, where food and drink was available. We enjoyed the sandwiches, tea and cake provided by the wonderful ladies at that venue, so much so that a trip to the TT in the coming years was not complete without visiting Barbara, Isabel, Kathy and Minnie.

We waited and waited for the race to start, updates were forthcoming regularly but no good news as yet. Brian was talking to the marshal telling him of our situation and was there another way to get to our car at Braddan Bridge. The marshal said he didn't think the roads were likely to open soon and so he'd allow us to walk along the footpath to Braddan Bridge on that short section of the track. We started the walk with his blessing and got within sight of the church on the second of the 'S' bends of Braddan but on the opposite side of the road when a big Pan European bike with a travelling marshal on board shouting at us to get off the track. We couldn't hear him properly but he left us in no doubt as to what he wanted us to do, The footpath ended before the church and another marshal beckoned over the road, we learned that day that before the race start the travelling marshals had a section of track to patrol for just such events. If they had guns I'm sure they'd shoot you for being where you shouldn't be on race day.

This presented us with another obstacle to our grand plan, we were now only a hundred metres or so away from the car but on the wrong side of the track so, taking advice again we made our way to The Quarter Bridge pub walking through fields and there we could cross under the road via stream that was negotiable and that would put us on the correct side of the road and we could walk over another field to the car. We walked to the pub and realised it was a good viewing point with advantages and decided to stay their 'til the racing was over, this time it was Castletown Ales. I should say that this happened long before the TT relief road was built and that it was the one and only time we went in a car.

We learned that it was best to book ferries and accommodation whilst we were there for the next year and so we did, staying at places like 'Mad Ken's' the Rio, but as the bikers called it 'The R10', The Rosemount, The Edelweiss, The Castle town Hotel in Castletown and of course our favourite for entertainment, The Cragge Hill Hotel on Hutchinson Square. Mary was the name of the owner, a short plump Irish lady much on the style of Mrs Brown without so much cursing, Alan, her husband, was ex-RAF but in a comical way but he did have his own light aircraft, more about that later. We were in the habit of getting the midnight crossing from Liverpool which got to Douglass at around five in the morning, a key would be left in a designated place for us to let ourselves in on arrival. Knowing our room numbers from previous calls we'd make our way to them with our bike gear and luggage. Mary would greet us about eight-ish for breakfast and we'd eat like food was going out of fashion , a full English with endless toast and tea. To say the Cragge Hill was a hotel really did flatter it, it was as near Fawlty Towers as you can get, but for a week with friends and new friends it was bearable....... just. After breakfast, I was rooming with Steve Cave, Brian and Pete, Steve mentioned to Mary that there was no dressing table in our room on the third floor. "You and Don come with me I'll sort you one out," she said in a Mrs Brown accent. I'm sure that if a camera had've been there she would've looked into it with a knowing smile and winked. She took us to the other end of the hotel and down to an enormous cellar which was full of very old hotel furniture. "Sort one out from this lot and take it up to your room for me, I'm too old for that sort of thing now," she said, and promptly left us to it. Steve and I sorted a dressing table and wrestled it up four flights of stairs, we couldn't stop laughing at what we were doing, us supposedly hotel guests.

I mentioned that Alan had a light aircraft he kept it at Ronaldsway airport on the island, it so happened that Pete had a pilot's licence and let it be known to Alan whom was serving behind the bar at the time. An indication of the hotel organisation is that he phoned up another hotel on the sea front to ask what they were charging for bottled beers whilst serving us. Still chatting he asked if we would like a trip around the island for twenty five quid each in his plane. We agreed and it was to be the next day, we met him at the airport wondering if this was such a good idea on such a windy day. Anyone that has been in a light aircraft knows what an experience it can be when affected by excessive turbulence. The take-off was a little bumpy but fine and we headed for the West side of the island, we found the coast and I'm guessing, at about two thousand feet and clear visibility. Alan changed our heading to North and it was fun picking out landmarks and the TT circuit. We soon were at the Northern tip of the island, Point of Ayr, and changed course again to a Southerly setting, although we could still see clearly the coastline. Alan said, "Do you want to take control of the plane and have a fly?" Alan asked Peter. "I've not flown for a while, but if you don't mind, I'll have the controls," said Peter, thanking him. Brian and I were sat behind the two pilots, listening, chatting and taking photos. "It's becoming misty Alan, shall I go down?" "Yes," answered Alan. Peter took us down to a thousand feet and levelled off. Brian and I taking more notice now. "Go lower," instructed Alan. "I can't fly lower than a thousand feet," Peter said. "Get down until you see the coastline NOW," an agitated Alan said. Brian and I nudged each other at the conversation but were anxious as Alan sounded worried and he took over the controls. We were pretty low by the time we turned in line with Ronaldsway runway and the landing was fine.

Organising the next trip Brian told us that he couldn't contact Mary and that he'd been trying for weeks. He did keep on trying and weeks later he contacted a tearful Mary who explained that she, Alan and their friends, a married couple, had taken a trip to France in Alan's plane. The plane had developed a fault on the return journey, and they had to land at Jersey in the Channel Islands for repairs. The repairs were to take longer than anticipated so arrangements were made for the ladies to get a flight home and the guys would make their way in the plane after repairs. Whatever happened was catastrophic because no mayday call was received, neither of the guys were seen again and only a small piece of wreckage was found washed up on the shore of the Calf of Man, a small island to the South of the Isle of Man.

Those trips took up the next twenty years, Tom didn't get a

bike and so didn't make it again for a good few years until a car load decided to come with us, Brian's dad George, Neil Johnson, Tom and my nephew Andrew Kemp in Andrew's XR3i. He got done for speeding in Kirk Michael and I went with him to the police station to pay the fine. There was a queue outside the station with at least thirty lads in it, all in bikers gear and carrying helmet and gloves. The queue soon went down and once inside it was easy to see why, there at a table were three people seated, the first one reading out the charge, the second asking 'guilty or not guilty' (I took him to be a magistrate) and the third taking the fine stipulated by the magistrate. They obviously had come across this type of thing before!. Although the island has no upper speed limit they are very keen during TT fortnight that bikers and others observe the speed restrictions in the towns and villages.

That reminds me of another tale about young lads, our regular crew involved big Bill Hutcheson, his son William, Tommy Logan and big Bill's brother, and that trip fell on William's eighteenth birthday. William had ridden pillion with his dad but had a full licence himself and was insured for his dad's 850cc Suzuki. On the morning of Williams birthday, a none race day, William had arranged to borrow his dad's bike and give one of the waitresses a tour around the racetrack, all went well and they returned safely. That night we were celebrating in the hotel bar and in walked a policeman and woman led by Mary, the policeman read out the bike registration and asked who owned it. Big Bill admitted he did. "You are charged with exceeding the speed limit at Parliament Square Ramsey, this morning at 10:30am," said the policewoman. Big Bill explained that he hadn't ridden his bike that day. "Then who was riding the bike then?" she asked. A cowed, red-faced William meekly admitted that it was him, all the exuberance was gone from the atmosphere in the room by now. "Then you will be required to pay a fine of, (Pause) A DANCE WITH ME," she shouted and immediately took off her uniform top exposing herself and grabbed a totally shocked William, forcing him to dance closely with her. Everyone was for a split second stunned and then erupted into laughter when they realised that it was a strip-a-gram, how big Bill and Tommy kept straight faces throughout the performance I really don't know but it set the night up to be an epic. Happily through those wonderful years not one of us were involved in an incident or accident. Lucky or what?

After the Isle of Man trips came Holland, Le Mans, Spain and Switzerland employing different routes each year taking in the sights and the night life in the stop-over towns, and each time the bikes got bigger. Another guy joined our group called Lance, a lorry driver with continental driving experience, we decided a trip to Holland taking in Amsterdam would be in order. Brian picked a hotel he'd been to on a company trip, and it was in Breuklyn, midway between Amsterdam and Utrecht, with a rail connection linking all three. After a day's touring and an evening meal we decided to go into Amsterdam on the train, our station being at the edge of the car park. Arriving at Amsterdam Central we immediately headed for Damm Square and hit the 'Brown Bars' enjoying the atmosphere, the company and the beer. Lance making contact with a short American guy that threatened to 'put a cap in his ass' if he didn't stop taking the p*ss out of him. Lance and Pete wanted to go to a show, Brian and I had had enough and were heading back to the hotel, so we parted company. Brian and I got on board our train and promptly fell asleep waiting for it to leave, only waking as it pulled into Utrecht station. Dazed and still half asleep we found a taxi rank and instructed the driver to take us to Hotel Breuklyn, costing ten guilders, around thirty pounds. On waking the next morning and getting ready for breakfast I said to Brian "We'll have a tale to tell them this morning over breakfast."

"We sure will," he answered. Lance and Pete were later than usual, but on arrival they looked the worse for wear. "What did you two get up to last night?" I asked. "You tell him Pete, I can't remember," said Lance. "Well it was like this.........." I'll tell the tale because anyone that knew Pete knows how long he could make a tale last. After they left Brian and I they went to a bar, had another drink and decided that they'd had enough too and made their way to the station. Studying the timetable they went to the wrong platform after buying their tickets and got on the wrong train and fell asleep en-route. Pete woke up to find the carriage in darkness and waking Lance "Where are we?" asked Pete. "How the f**k should I know," snapped Lance, still dazed. They forced open the doors of the carriage and realised just how far to the ground it is without a platform in situ. As they studied their situation a rail employee came along on his inspection, got them on Terra firma and took them to his hut. The guy apologised profusely and told them that they were at Lillistadt, the end of the Northern Line, in railway terms. He made a phone call to his superior that gave permission for a taxi to be hired to take them back to Hotel Breuklyn. It transpired that it was their fault for parking the carriages for the night without inspecting them properly and the guy had seen emergency lights come on as Pete and Lance opened the carriage doors. We really can't be trusted to go out at night alone!

Pete didn't forget that kindness and within a few months on his way to Switzerland he took a bottle of spirits for the guy that helped Lance and he that night.

There was a couple of years on our Swiss trips when I couldn't ride, one when I'd had a carpal tunnel operation and another when my bike was in for fairing repairs. It was to become a standing joke amongst our spouses that we were going on a bike trip, in a car. Those trips were memorable in that we were not used to such comfort and could share the driving and so could travel further. One trip took us South to Millau in France to see a terrific viaduct, so high that some days you drove through clouds (We were to repeat that trip on our bikes a few years later) and other places and the second to Monaco. On our way through France we stayed in a hotel in Reims, I was driving my 'S' Type Jaguar 3L V6 SE and couldn't find a parking space, so I parked in the exit drive leaving enough room for traffic. We checked in and went to the room allocated to us. Looking out of the window at the car I could see a vacant space. "I'm going to move the car, I'm not happy with where it is". I told Pete. I went, moved the car and returned. "Bloody hell!. That's service for you. The reception phoned whilst you were gone and said your car was being stolen. I told them what was happening, and they said that they'd kept an eye on it as it was such 'a beautiful car'. "I'm getting a Jag if it gets attention like that!" Peter told me.

Millau viaduct

'Born Lucky' Book Two: From Mynah to Miner and Beyond

The 'bird

Beatenberg

'Born Lucky' Book Two: From Mynah to Miner and Beyond

Amazing place

Gelmerbahn

97

'Born Lucky' Book Two: From Mynah to Miner and Beyond

North face of the Eiger from our hotel

On the bike in the Alps

On arrival in Monaco we did the usual sights of the casino, wondering at the opulence, the yachts and the high up the hillside mansions but the biggest thrill for me was to drive around the roads that the Grands Prix is raced on. I was totally overawed by that track. We left Monaco and headed for Switzerland stopping overnight in the city of Cuneo heading towards Turin. I parked outside the Best Western Hotel, and we got a room, checked in and went to the cafe next door for a beer, having left the car keys for them to put my car into their secure parking area. We sat taking in the city square, culture, and people watching as we always loved to do everywhere we travelled; when a hotel employee came to us with the car key "Ah, I know where the casino is, I have the night off and now I have the car," he said with a heavy Italian accent. I asked if he had enough money for an evening in Monaco. "Sadly not," he said getting into my car and driving it away. "I'm definitely getting one of them," Pete said as we signalled for another beer.

From Cuneo we made our way past the Matterhorn towards Mont Blanc and to Chamonix then Switzerland onto Kandersteg in Switzerland where we travelled with the car on a train through a mountain tunnel to Frutigen, that put us onto the Susten pass, a favourite of ours. Then on to Interlaken up the Niederhorn mountain about eight kilometres to Beatenberg and there is the very welcoming Alphotel Eiger with Chris and Katrin Ringennberg with their son Marco out front to greet us as always.

Brilliant roads on a bike

We spent many happy hours biking to Switzerland via many different routes through France, Holland, Belgium, Luxembourg and Germany, but touring through the high passes in France, Switzerland and Italy has to be the pinnacle of those trips, even doing the Stelvio Pass in Italy. That pass has forty-eight hairpin bends and for thirty six of them you rise through a thousand metres up the mountainside, that tests your bike handling I can tell you, but my trusty Super Blackbird 1100XX never faltered bless her. I treasure those times, the visit to the Schilthorn revolving restaurant (seen in the James Bond film 'On Her Majesty's Secret Service') via those terrific cable cars. The Reichenbach falls where Sherlock Holmes met his fate, there is a statue of him at Meiringham, a nearby village. The frightening trip up the Gelmerbhan on little more than a direct haulage in an open carriage on rails.

The meals at Ruigli Bruigli overhanging a mountain edge and overlooking the western lake of Interlaken, the Brunersee and the wonderful views at breakfast time of the valley, the lakes and the North face of the Eiger, The Monck and the Jungfrau mountains.

Brian and I ventured to Le Mans one year for the twenty four hour car race, we rode to Portsmouth, met up with some guys that Brian knew, four more on bikes and others with vintage Le Mans Bentleys, open topped and other marvellous old cars. As we crossed to Cherbourg we enjoyed a couple of beers in the sunshine on the ferry as we were staying overnight in a Cherbourg hotel. We enjoyed rather more beer that night, we missed breakfast and were only awakened by the smell of petrol and a gigantic roar as the two Bentleys were fired up in the underground garage, we quickly dressed, settled our bill and left.

We weren't following them as we'd decided to travel via a national park to the camp site at a village called Beaumont-sur-Sarth which was to be our base. Tent pitched and secured we were due at a local restaurant for 'a sea food banquet' apparently something these guys did every year. We were mixing with bankers, financiers, hedge fund managers and the like, it was no bother at all, we were new to this scene, but they welcomed us with open arms and it was great. That feast was something else all the shellfish you could name, Cockles,

Mussels, Whelks, Crabs, Lobsters, sea bass, herrings amongst many local delicacies and with all of the implements to tackle them with. All were presented immaculately by a Maitre'd called Mercedes. Brian mentioned to me that it would be advantageous if we could get a lift to the circuit the next day and set me up to get one. "You've got more nerve than me." A ploy he'd used many times in the past. I put on my best confident voice and asked the first guy to come near us "I say, Neville, how are you getting to the circuit tomorrow?" "By car Don, would you like a lift in, were booked in car park Rouge right near the entrance." "My word, that's decent of you that would be great, thank you." We arranged a time and Brian said whispering, "I don't know how you've got the nerve, you p*ss taking b**tard."

Next morning we met up with Neville and John and they took us to the car, it was a Volvo, one of those sporty ones with the flip up headlights and we made off to the track. If the track was thirty miles away we travelled about fifteen of them before the car lost power, didn't stall but wouldn't pull well. Everyone got out and Neville made a call to his recovery contact who said it would be an hour. That was no good to us, the race would be starting then at three o'clock. Neville decided to risk it and we set off again, very slowly at first but we did maintain a respectable thirty miles per hour on the flat, although you wouldn't think so, hearing the horns blaring from mainly English cars. We managed to get within a mile of the track and the traffic was so slow that we decided to get out and push the stricken Volvo into car park rouge.

Once there we headed for the main gate and once there we heard the mighty roar of engines, "We've missed the start," John said to no one's surprise. Passing through the first set of security and not yet joining the paying queue I stopped to get some cash out, dropping slightly behind the others. Looking up I made eye contact with a massive black guy, he said, "Hey buddy do you want this, I only wanted to see the start," and thrust a ticket into my hand. I thanked him for his generosity and made my way to Brian and told him what had happened. "Do you think it's real?" he asked. "Nothing to lose, if it's fake I'll just come back and pay," I answered with a knowing grin. We marched to the ticket collectors and through I went to the Le Mans racetrack, permit all areas. "I've never met anyone as jammy as you," he said.

We enjoyed a fantastic twenty-four-hour race, moving around the track, seeing flames and glowing brake discs in the dark hours at the end of the Arnage straight, even leaving the track for a three course meal. Having eaten the three courses a waiter brought me another pudding by mistake, but Brian shamed me into giving it to him.

After the race there was a mass exodus of the British contingent, that's a sight to see, mile after mile of nothing but

British registered cars and the French locals waving at them from the roadside and from bridges. Brian and I returned to an almost empty campsite, we were there for the week or at least in France for the week. We rested and showered and just before nine we made our way to the local bar for a bite to eat and a few beers. SHOCK HORROR!. It was closed, I knocked the door and the lady explained that they only opened late when the race was on, the rest of the time they closed up at nine o'clock. Disappointed we made our way back to the tent and drank a litre of J and B whisky we kept in reserve for just such emergencies. The rest of the week we toured taking in Pithiviers and Bayeax.

Rest in peace
Big Bill, Tommy, Brian and Peter.
Keep the bikes warmed up and ticking over,
until we meet again old friends.

'Born Lucky' Book Two: From Mynah to Miner and Beyond

Hardships and Problems

In my capacity as Colliery Overman I would arrive at the mine about five thirty, always walking through the canteen as the night shift workmen would be in there having breakfast or just a cup of tea and a cigarette before leaving the mine. That was always a productive few minutes, men would beckon me and tell me of their hardships through the night or about a long term problem they were having. It was generally a precursor for what the night shift Colliery Overman would be telling me, although Johnny Edwards would be miffed at me for keep saying 'I know' at each snippet of information that he imparted to me, so much so that at times of pressure he'd say, "For f**ks sake Don, have you been here all F**king night, you know more than me!"

It was one such morning I walked through the canteen as normal and Wilf Robinson and Bill Crowson caught my eye and gestured to me to join them, this was out of the ordinary as just shouting information was the norm. So deciding to get a cup of tea, served by the lovely Iris Blakemore and Madge Smith, I made my way to Wilf and Bill. "Hell fire Don, you want to see the state of 68's panzer (The armoured flexible conveyor that carried coal along the face and upon which the powerful coal cutting machine travelled) it's stuck fast," said

Wilf, a tall rangily built man, I liked Wilf. "We've been taking off plant pots with the fitters (Extensions to the face hydraulic supports legs) because they were too much and were almost making the chocks solid between roof and floor and potentially stopping the face," said Bill, of similar build to Wilf but unshaven. These were formidable workmen, and you'd always pick them in your team. "Anyway". Wilf carried on. "We'd had a good shift and completed the main gate half of the face, putting the unwanted plant pots on the panzer, to be run off when we'd finished our shift." Bill then continued, "We let the panzer go thinking there was a man watching the delivery onto the stage loader with such a potentially hazardous load on it. Everything was OK for a short time but then we heard high pitched scraping and banging noises and so stopped the panzer, locking it out (a safety device placed every seven metres along the face line)." Bill explained. Wilf, eager to tell me, said, "Some had got stuck at the delivery, piled up and were caught by the bottom chain and were taken along the bottom of the pans where the gap is around 10cm, but the width of the plant pots is around 20cm diameter. They've bowed the deck plates upwards and busted up some of the inspection covers (a place where the bottom chain can be inspected at intervals along the face length) and the panzer is stuck fast. The fitters are at it now. With that I knew what I had to do, telling Tony Prime to deploy the day shift men for me, I got changed into my pit clothes picked up my rescuer, cap lamp, safety lamp and tallies and headed for No 2 pit top. Telling Johnny Edwards to send the first six men of 68's team to follow me in the pit asap.

On arrival I found two night shift fitters, Spike (don't think I ever knew his name) and Alan 'Cracky' Watts finishing the taking off of the inspection plates to where they thought the first of the plant pots had got to, it was almost halfway along the 220m face. "What do you think Spike?" "Don't know Don, never seen anything like this before." I liked Spike and Cracky, good 'get you going' fitters but this was a first for me too. I'd experienced all kinds of fast panzers, carry back, packed races, staked, falls of ground and broken chains but not this. "We're leaving now there's nothing more that we can do," Spike said. "As you pass the control panels, isolate the loader, put it in 'contactor'. Put both panzer motors in reverse and let me know when you're ready," I told them. A few minutes later they shouted on the tannoy that they were ready. I was now at the main drive. "Let it go in reverse, I'll stop it Spike." "Righto," came back the reply.

The pre-start alarm sounded, and I was in a safe position with my hand on the lockout button. The panzer started, I let it run for about five metres, there was some terrific banging, and a few plant pots came out from the bottom race at the delivery of the panzer. "Let it go again," I ordered. I was used to the unholy scraping and banging now with no apparent damage, the panzer started, I let it go twenty metres this time. More terrifying banging and scraping plus a fair few plant pots this time. "Hold it Spike, I need men here before we move it anymore. Put the panels in neutral and make your way out, as you pass the day shift hurry them up." "OK Don, will do." With that I came off the face to phone the under manager Harry Gibson, a clear thinking Geordie whom I got on with very well. "What do you think Don?" "I reckon with a following wind we'll be going by ten o'clock." "You've shocked me, all the estimates up here range from three shifts to three days, are you being too optimistic? I daren't tell the gaffer three hours and be wildly out, he'll go into orbit." "Tell him what you need to but I'm sticking to three hours with a following wind, the day shifts here now, got to go."

Ron Wileman, Nicky Ross, Neil Chapman, Melvin Green, Roy Reid and Mick Burton had been sent to me by Tony Prime, good, I had a machine team as I'd told him and now I could prove myself right. The fitters were later as always but it was heart-warming to see Tommy and Roy Jacobs the two best engineers I thought. Speaking to them I told that I would post a man at each inspection cover to stop the panzer and remove any plant pots as they appeared with the panzer in reverse, plus a man at the delivery end doing similar duties. "That's great Don, as you're doing that we'll assess the damage to the pans to see if we need to change any, there's some being loaded on the bank as we speak," Tommy said.

Reversing the panzer alarmed Tommy and Roy with the racket it produced but they let me have my way. It wasn't straight forward as plant pots popped from the bottom chain through the inspection aperture they would get fast on the top chain and either jamming the panzer or flying in any direction, very dangerous. Roy came to me and told me that some of the inspection covers could not be refitted as a piece that the retaining clip relied on had been ripped off by the upward pressure exerted by the plant pots. "Is that a stopper Roy?" I asked worriedly. "No, I've arranged for drills, taps and countersunk studs to be rushed in by a runner (a fast moving young lad) they'll be here in a half hour." "I think the pans are damaged but they're serviceable and we can change them as an organised job at night and weekends," I said to Roy hoping he would agree. "That's exactly what Tom and I have said Don," Roy agreed. I was still on to meet my goal!

One plant pot had got further than we thought but after a struggle we got it free, Roy and Tommy's drilling and tapping took longer than expected and we had to place a man at those inspection covers in case they came adrift whilst coaling, but we were winning coal at ten thirty. Not a bad day all in all, just shows the value of good information, knowledge, faith in the people around you and teamwork.

Work on Day shift was always more intense than other shifts, there were more bosses with their own particular problems and when they needed help, who you gonna call..............Don Povey. The job was intense, demanding and required knowledge, skill and command of all the calm and mindfulness you had in your armoury. I was fortunate that when I left the mine that I could close my mind to it, especially after thrashing that 650cc Kawasaki SR to within an inch of its life. I did enjoy those rides home.

"Hello, who's that?" said Johnny 'Acker' Adkin, answering the telephone in the deputies underground cabin with one foot on the chair beneath it as everyone did. "Don it's for you, it's Cavey at 65's." "Hello Steve, what's wrong?" It was force of habit to say 'what's wrong', rarely did anyone ring to give you good news. "Don you better come quick to 65;s return junction, it's sitting down before my eyes, I've never seen anything like it." "Don't panic Steve, keep everyone away from it, with a fence if necessary and you guard it for safety reasons. I'll be with you in fifteen minutes." With that I set off, Mr Betteridge coming with me, he had overheard the conversation. Ian was another fine under manager, practical and understanding. I shouted to George Botterill "Get some chock wood in the pit George, I've a feeling I'll need i.t. "There's some coming up Power's (Arthur) drift now Don,

Alan Fantom and Rodger Roberts have come on early and helping out," he answered.

Making our way across Zahan's slit quickly we soon ate up the metres to Cavey whom fortunately had been on the outbye side of the junction, he was snapping but got straight up as we arrived. "No cracking, banging or groaning Don, it slowly sank," said Steve. "OK," I mumbled as I weighed up this disaster. 65's junction held the key to supplying three faces on that side of the mine as well as a service access to the belts and 65's bunker. Cavey was now eating his apple, must have been a juicy one, I've never heard such acoustic enjoyment since I gave an apple to Star, my pony at the time. It was affecting my concentration; such was the level of pleasure coming from Steve's lips. "For f**ks sake Cavey stop eating that apple." He was shocked at my outburst and threw it away immediately, well, I couldn't think could I! Mr Betteridge laughed.

I estimated that the junction had sunk a good 60cm into the floor uniformly at the four corners and it was only one point five metres previously. We had had problems the before, some months previous but nothing on such a scale. I immediately sent for 'Chippy and Morton' of syphon fame in my previous best seller. "Bring your tools and get to 65's junction, if you crack on there's three loads of chock blocks leaving the pit bottom soon, Colin Fern will bring them in for you." Also redeploying Maurice Hannigan and 'Lonza' Freeman to the site. Two old heads at situations like this. I told all four what was required and Mr Betteridge agreed. I needed both carriers, a pair each side, bolstering with chock wood. Although this would be a temporary inconvenience, it would allow us time to dint the track and lower it for reasons of load clearance.

It took a couple of weeks to complete that job, it caused disruption and hardship but nothing that you wouldn't expect underground from time to time. Steve and I often recount the story of me exploding and telling him off for eating his apple.

A mine deputy is required to read the barometer each shift that he attends. This is to give him an indication as to what to expect regarding gas emission during his shift, if the barometric reading is low or falling the likelihood is that emissions could be more prevalent. That reading must be recorded in his statutory report. Mine gasses are termed 'Damps' by the mining fraternity and there are many, Firedamp (Methane), Black-damp (Carbon dioxide), White damp (Carbon monoxide), Stink damp (Sulphurated hydrogen) to name the more common ones.

There are many places an experienced deputy would search for them, firedamp is lighter that air and so could gather in roof cavities or be issuing from a 'feeder' in or above the coal seam. Black-damp is heavier than air and so could be found in swilleys or issuing from stopped off abandoned workings.

White damp is almost the same specific gravity as air and is found in the 'general body' of air, mainly where spontaneous combustion is taking place, that happens a lot in coal mines. Stink damp is extremely noxious and smells of rotting cabbage, the others being colourless and odourless (The gas in your home is methane but has a smell added to it by your provider) and is found in such places as water lodges for mine drainage.

Cue Ken 'Tat' Tattersall phoning the deputies pit bottom cabin from 69's main gate stable hole "Is Don there Acker (Johnny Adkin)?". "For you Don". Said Acker, offering the handset to me. "Hiya Tat what's wrong?" Tat wasn't a guy that phoned very often mainly because he was a bit shy but also he could solve most problems himself. "I've found gas in the stable hole before I rammed the shots". Meaning that he'd done his duty and tested for methane gas before charging the shot holes with explosives and detonators. "OK Tat, I'm on my way, Get Rod Noon to bring his methanometer from the supply gate to you, to make sure there isn't a fault with yours." "OK Don," said Tat, sounding annoyed that he hadn't thought of it.

Arriving at 69's main gate I immediately went into the stable hole, I noticed that the machine was bearing down on the stable from about halfway, Billy Ramsell not taking any prisoners with his shearer, the panzer was heaped with coal.

Those shots needed firing to avoid any delay and there was one and a quarter percent of methane present at the mouth of the shot hole. Mining law states that at one and a quarter percent if found in the general body of air flow, shot firing shall cease and electricity switched off. At two percent men shall be withdrawn.

Deputy test for gas

'Born Lucky' Book Two: From Mynah to Miner and Beyond

Me on the coal face

Coal face

Typical underground explosives

Although this was not in the general body and we could legally fire the shots, to allay the fears of the stable men, Geoff Bunce and Jeff Harris, I decided to erect a 'diverter' shield made of lagging boards to get a positive current of air into the area. That lowered the percentage to a quarter percent, and they were happy. Everyone was apprehensive and crouched low in the side of the roadway behind the chain mail curtain and the steel protective shield as Tat shouted 'Fire' and pressed the button on his ME12 exploder, but all went well.

Ellington(Northumberland)

Charging a shot hole

That procedure was adopted for the rest of the week and over the weekend as the stable had fallen behind, although being modified by using a custom-made air diverter. We weren't a gassy mine, indeed before nationalisation Ellistown was a 'naked light' mine in which smoking was permitted; so any indication of methane set alarm bells ringing and made men, officials and management very wary even though the regulations are very clear.

'Born Lucky' Book Two: From Mynah to Miner and Beyond

A Change of Scenery

The kids were growing up now Andrea being ten and Stuart eight, we enjoyed the usual family things like going the Ferrers pub on a summer night, trips to fun parks and swimming, everything that families with young children enjoy. Our annual trip to the Blackpool illuminations was a highlight away from the normal summer holiday in Spain and Cornwall. The kids loved the funfair but loved the slot machines more, overjoyed when they'd won fifty pence but took some convincing them that they'd made a loss by spending three pounds to win it made no sense to them at all.

Stuart had joined a junior's football team and the coach, Pat Carty collected him for initial appraisal and training. Sue and I had often sat in our lounge on a Sunday morning looking outside at the foul weather. observing Pam and John Birt loading their car with the family and going off to watch their son Simon play football. Sue would say "I couldn't do that, out in all weathers, getting wet and freezing cold every Sunday, I enjoy my comforts too much." We were soon enveloped fully into Stuart's football career albeit summertime and the weather good, we knew what was coming and Pam and John laughed.

I had talked with Sue for almost ten years about the expected life of local mines and that to maintain our comfortable lifestyle we'd probably have to move nearer to Melton Mowbray. At first it seemed a lifetime away, but time passes quickly and decision time was upon us. A position was advertised at the mine for Asfordby, after much thought and with no advice I applied for and succeeded in getting the job of overman at Asfordby. Sue was in tears as we left Huntingdon Road and all our close friends and neighbours that we had spent the last fourteen years with, it was hard for her.

We settled in Melton Mowbray, I had my work, Sue had a little job at the kids' school and Stuart was soon picked the school football team and joined Melton Foxes football team too. Andrea quickly made friends and within weeks she was holding sleepovers with Emma Farrell, they are friends to this day. Those football Sundays made us some lifelong friends, Paul 'Brasso' Knight and Karen his wife, Phil and Val Curtis, Baz Hatton, Bruce Forrester and his wife to name a few. Sue soon took over the running of the 'subs' book and collected the money weekly from the mums and dads. She wasn't shy about going into the changing room of those ten-year-old boys either to get her subs, loud shouting and protestations could be heard from them as she entered, but she got her money.

I had promised Andrea and Stuart that they could have a puppy when we moved as something to look forward to. Months passed, I had redecorated throughout this five-bedroomed house, knocked through from lounge to dining room and redesigned and dug the garden. "When can we have the puppy you promised us dad?" Andrea asked. They had both been terrific, so the next night we went and bought a Cavalier King Charles spaniel from a local kennels. We were discussing the names for our new family member a bitch, travelling home in the car and several were blurted out then Andrea said "Penny". I immediately replied, "Not likely when I've just forked out a hundred and fifty quid for her." (1987) We called her Tess by mutual agreement, but I can't remember whose suggestion it was, she was to become the centre of attention for four fourteen years and kept up after that by her daughter Tammy.

I enjoyed my pint on a Friday and Sunday nights at the White Hart in Melton, both nights there were guys playing darts, the board being over the fireplace and the throw parallel to the bar but away from it. The bar was a high one and Paddy Lohan looked down on you whilst serving. My favourite barmaid at the time was Edna Kettel, an older woman with a wicked tongue on her when upset and of course the guys tried to rattle her cage. She was a biggish woman and behind that raised bar she appeared massive.

There used to be a little Cornish man come in on Fridays and as soon as Edna saw him she'd scowl but out of his view, every time you caught her eye and he wasn't looking she'd make a face of scorn. He'd order a half pint of Marston's Pedigree, raise his glass to the light, then smell the beer and sip it. "Not quite as it should be Edna." "Take it or leave it," she'd snap. This went on for three half pints and by the fourth she'd be seething under her breath; he never moved from the bar or asked his beer to be changed. "Good night Edna and thank you." "F**k off you little b**tard, you'll get this the next time you come in." She was waving aloft a policeman's truncheon she kept behind the bar for such occasions. Her husband was from London named John Moulds and was a lovely quiet man that played darts with us. He got those looks sometimes too.

The Sunday nights were quieter and there would around five or six of us playing and I'd be doing OK, Carl Wilson, a Scot came up to me. "Don, have you ever played for a team?" "I captained my team back in Ashby Carl," I answered. "Will you play for us tomorrow night?" he asked. "I'd like to get involved in Melton darts," I answered with a grin and a handshake. Sue enjoyed those dart nights along with other ladies as she had done in Ashby, all the Melton pubs paid the league subscriptions and put on a hot supper. We were in the second division and doing OK but not good enough for promotion. I played with Carl Wilson, Pete Tunstall, Terry Rate, Pete Haynes, Joe Maguire, Terry Rate and his mate Richard for a few years and it was great.

The 'Irish Embassy' as the White Hart was known, changed hands and Jim Robinson moved in, a family man with two daughters, Jane and Sue, they were great behind the bar, chatty, witty, quick to serve and if they were busy they always made eye contact to let you know that you weren't being ignored. Jim was always doing something at the pub, it was great to seeing it being updated. He even started an annual darts competition; it was called The Tom Sweeney Shield. Old Tom used to stand at the bar and watch us play for hours on end and he donated this shield before he passed.

The format being 501 double in double out, best of three. The ladies could have either 401 double start or 501 straight in. I won that trophy three times. The first time I took the family with me, and they watched as I beat 'Duke' Ellington, the match going one apiece and the third leg I was on double top before he got off, then he got off and I ripped double top down to double one whilst he was coming down a ton at a time but I nailed that double as he left himself tops. Tom's wife Margaret presented me with trophy that night. Jim moved on after a few years and the pub lost its custom and closed but Jim came back to Melton to run The Boat a few years later. The team kept together, and we played from the Rutland Arms and then the Generous Britain for a few seasons but by then I had to go on shifts at work and the team folded, it was a shame because we'd won our division and got promotion to the Premier League.

My shift pattern changed but I hadn't a team to play for but luckily enough Stuart and his mates, all eighteen by now had formed a team and were entered in the second division from the Generous Britain pub. I was keen to help them, so much so that I used to do the fifty-mile trip from Daw Mill to Melton in forty five minutes when on afternoons to get there for eight thirty, match start time. They were a marvellous set of lads, keen, skilful and accurate but were not so good at reckoning, which nobody is when first starting out, and tactics, that's where I came in. They did very well coming runners up in the league and in the divisional cup. I really enjoyed that season with Stuart, Mat 'Cat' Catton, Paul 'Scouse' Stevenson, Gav Baxter, Wayne Czervoski, 'Bungle' and Chris 'Whiskers' Jones.

I played darts for the White Hart, The Generous Britain, The Rutland, The Black Swan, The Mash Tub and The Boat winning the Premier League four times with Ron Graham as Captain, Keith Tomblin, Tim Graham, Miles Hewitt, Murray MacInnes, Paul 'Scouse' Stevenson, Clive Rodrigues and Dave Hebb. The format being six a side, three pairs, six singles, and two triples. Each game won scoring a point and two bonus points for winning the match. I did check out on 170 at the

Black Swan during our time there. There is a league in the Vale of Belvoir, they play on Friday nights, and I was asked by Barney Kettel to play for the Castle at Eaton, we won that League five times, three from the Castle and twice from the Kings Arms at Scalford. When I first played in that league they played for beer each game and if you won your games you could be three pints in as well as in a round with your team, we were nearly always legless. Luckily big Dave Pearson didn't drink on those nights and drove us home. Those teams were generally the same as the Melton teams and the Vale of Belvoir teams took umbrage when we beat them, in a good way though. I made some good friends in both of those leagues and still have to this day.

Clive Rodrigues with our season trophies

The hub of darts in Melton Mowbray I've always thought in my time is The Half Moon in the market square, always friendly, always sporting and always a good game of darts. The landlord, Martin Davis fosters this culture by his own example, at the end of Monday night darts everyone congregated there for the last pint. I introduced a new player to the Melton Darts League, John Butler, a professional golfer. He was amazed at the camaraderie even between teams let alone within teams and soon made a lot of new friends, he was a good 'chucker' too. At that time the Love's played for the 'Moon namely Benny, the dad and John, Danny and Andy, a lovely family of Scottish descent. Danny told me that things were tough for him at the time and his wife Sinèad had complained of being under pressure making ends meet "Pressure, pressure! You don't know what pressure is until you've played The Boat at darts." Danny also told me after a particularly tough match that I beat him in "When you're gone you old c**t, I'll throw darts at your headstone." "You'd miss," was my reply and he'd shake my hand and hug me, laughing like anything.

Ron and I winning ways

Walking home one night after leaving the 'Moon with 'Golfer' John Butler, he would stay overnight with me on darts night and leave his car in town; we passed through Snow Hill an industrial estate. Half cut and glancing between buildings, we saw a guy dressed in a proper spiderman outfit and carrying a briefcase. There were three others with him next to a fire escape. "John, John, leave them alone," I pleaded, John is not that sort of guy. He approached them, shook hands with the convincing Peter Parker guy, picked him up and swung him around and on putting him down almost fell over. He asked repeatedly to shoot out his spiderweb from his wrist unit and generally made a nuisance of himself. They were students filming as a project and this was there location. John tells that tale often and I have photographs of that exchange.

John with 'Spiderman'

At the mine sinking operations were underway and all was going to plan, I found myself reporting on such things as sub-station building with rounded corners. Ditches, water settling construction, temporary winding tower erection and winder installation, both shaft collar formation, a Social Security raid of the workforce on site to check their credentials, the wheelwash, air sampling points etc. etc. It was a massive undertaking and difficult to report on with any degree of accuracy when your knowledge of the building industry and terms is little more than zero, but what I did know about was safety and that gave me a good footing, safe use of tools, clear walkways, helmets, compressors and their baggings (hoses) and that was my main concern.

The shaft operation ant the upcast (Where the mine ventilation flows up to the surface) was back wall grouting, where a liquid grout is injected behind the concrete lining of the shaft. It serves two purposes, to even the side pressures on the shaft lining and to seal any cracks in the shaft lining whilst the concrete is curing, thus stopping any ingress of water.

It was a Sunday morning and there had been a heavy overnight frost, the upcast shaft was grouting, the downcast shaft was stood down for maintenance that shift. I heard a shout from Paul Kaston in his broad Liverpudlian accent, "Where's the water Don? Send some f**king water. How can we grout without without f**king water?" He was asking Elsdon Ward at the pit top. "Sorry Paul, it's all frozen up, we're trying to thaw it as best we can," Don replied in his softly spoken Welsh tones with no swearing. "Let paddy Know that I'm bringing the sinkers out of the shaft to help, we can do nothing constructive down here." "He's here and he's heard you, he says come up here." With that Paul comes out of the shaft with his sinkers and begin tracing the water pipeline looking for bursts and frozen pipes, not a straight forward job as there were many inter-connections along its length.

Whilst they were doing that Frank Parker had been towards the office block and reported back that the water tank wasn't frozen that fed the offices. I let Paddy know and he could have use of it if it was of any use, office workers don't know there's seven days in a week! "I'll get the sludge gulper (The kind you see cleaning drains at the roadside) and make my way to the tank," he said. With the sludge gulper parked on the slope and backed up to the tank, we removed the top cover of the gulper and started to bucket water into it from the office water tank, such was the urgency to get the shaft working. The three of us were soaked from spillage in no time when Robbie Cizmazia came running to Paddy. "We've got water to the shaft Paddy, it was frozen in the fields up to the Lancashire boiler tanks." "Thank god for that Robbie, thanks for letting me know." With that, we ceased our operation, Frank took the gulper back, Paddy made for his office, and I made for the deputies cabin where there was always a brew on and it was always warm.

I let the gaffer know that we were away and grouting again and Mr Skelding said, "Thanks Don, don't phone again, I'm out to lunch with the Mrs." It was a totally different Mick Skelding phone call to those that I'd had whilst he was the manager of a producing mine. They thought it was all over, but it wasn't. The frozen pipeline in the fields had been freed using acetylene torches, the pipeline being raised on supports to a height of 1.2m from the ground for about 150m. So to prevent them from re-freezing through Sunday night braziers were spaced appropriately along the raised length, being replenished throughout that shift. Someone over replenished them in the hope of saving a trip over the field with more wood, the flames reaching such a height that the hedgerow some four metres away of tall Hawthorns, caught fire and spread to such an extent that the local fire brigade had to be called to attend. The Monday morning was very interesting to say the least, British Coal didn't like to be in the local paper for the wrong reasons.

Another time, we hit the news media front was, again on a Sunday morning on re-entry into the upcast shaft in the early days of sinking. At that time we were using Cryderman loaders to load out the debris after blasting, one each side of the shaft. Production had not started as this was a maintenance shift, but two men had been sent to lower both Cryderman's into position and anchor the to the concrete 'tubbing' (circular steel concrete shutter). The Crydermans had to raised before blasting for their protection as did the 4m high working stage. The stage is then lowered to the working level for mucking out and the Crydermans lowered individually through steel trapdoors in the stage. This was being carried out by two sinkers in conjunction with the surface who were working the Cryderman hoists. The first one was entered into the top of the stage and being watched through the top two apertures, Tony South, the deputy was assisting as he always did, very conscientious was Tony, so the second sinker took advantage of this and went to the last trapdoor below and opened it. Tony heard a groan and looked for that second sinker, he was nowhere to be seen. Going down a level Tony could see the man on the sump floor some 5m below and lying on the blasted sump floor.

Eimco overhead loader shovel in the shaft

Cryderman loader in the shaft

Tony shouted for me on the tannoy for assistance, I immediately rang for an ambulance and asked Tony, "Should I come in the shaft to you?" "Not yet Don, got to make it safe to land a kibble first, Get the Neil Robertson ready and bring that and some splints with you when we've levelled the muck pile enough to land a kibble." "OK Tone, I'll sort it." A good reliable man was Tony. Whilst Tony was sorting the sump out I got the special stretcher and splints Tony asked for just as the ambulance appeared at the pit top, it was quick because Melton had its own station at that time. Seconds later, it was followed by the fire engine and a police car, the fire officer asking of what assistance he could be. "I would say that you can't help at all," I answered as I showed him over the shaft doors how far down the shaft was. "We'll be on standby if you need us to be of assistance." "OK, but I can't let you into the shaft to do any work, you're not trained," I said. Tony now had the Neil Robertson stretcher and splints, the sinker was responsive but shaken and dazed. With the various signals I knew what was happening below, they had fitted the stretcher to the sinker, detached the winding rope from the kibble, lowered it to the stretcher and attached it to the head end. They had raised the man in the stretcher into the kibble and re-attached the winding rope to the kibble and then ensued a very slow journey to the surface.

The phone was ringing before I got to the deputies cabin to ring the gaffer. "Don, Don what's happened?" It was a very worried Mr Skelding. "I've got radio Leicester on and they're saying that a man has fallen down the shaft, whose told them that and how come they know before me?" "When I rang for an ambulance I told then a man had fallen in the shaft not down the shaft, anyway Mr Skelding panic over, he's gone to hospital, in my view he's OK but Tony dealt with the situation excellently, no panic no shouting and no hullabaloo." "I'll have to let HMI (Her Majesty's Inspector of mines) know and I'm on my way to you." "OK Mr Skelding," I said.

Apparently when a 999 call goes in all three emergency services are notified along with the local radio station, hence it was broadcast before the gaffer had been notified. The rest of that Sunday was spent alongside 'Hannibal' Hayes, our nick-name for HMI, he could be vicious, but not on this occasion, and a full safety inspection team. As we entered the shaft for a site inspection Hannibal said to me "I can think of better ways to spend a Sunday afternoon than this Don, can't you?". "I sure can Mr Hayes," I answered. The sinker was at work next morning after a check-up at hospital, but we were in the local press again for the wrong reason.

'Born Lucky' Book Two: From Mynah to Miner and Beyond

Rewarding Times and Men

I was so proud to have got the overmans position at Asfordby, but initially there was nothing that I understood and it was a very steep learning curve where for a short period the deputies were more familiar than I was with the many firms, sites, working points and personnel of British Coal, quantity surveyors Bapte and Cementation, the main contractor. It only took a short time and I'm a quick learner and I used the time to get to know the deputies better, the office staff, the canteen ladies, the cleaners whom I had to let in the premises at night and lock up after they'd finished, a lot of those cleaners joined the permanent canteen staff after it was commissioned.

Sinking was done and the insets formed, now we were going horizontal, that's more like it, it was what I was used to, except to say it was more than twice the size and that brought bigger problems. Everything was more than twice as heavy and couldn't be manhandled, everything was twice as high and couldn't be reached without scaffold which underground isn't safe and can't really be made safe easily and certainly not quickly, also in the circular drivages there is very little floor space to put anything, even so I was now in my element, and I could influence things more.

The first British Coal men came and did specialist installation work, the like of which they'd never seen, and me neither but I had the advantage of seeing the plans before they came and knowing the pit layout underground and on the surface. They took to it with a relish and surprised even me after they had got used to reading and following blueprints. Platforms were being built, conveyor drive units and tension boxes installed, and conveyor structure hung in circular concrete roadways after drilling for its suspension points from scaffolding, (Rickety). They astounded me with their enthusiasm, willingness to learn, overcoming there initial doubt to achieve and their honesty. To a man they didn't say that they couldn't do it, they doubted they couldn't do it quickly, but I told them to do it safely, speed is desirable but safety is essential. Little did they know what I had in store for them, they'd need all those qualities and more.

Those that I had chosen to go in seam did just that at Q junction, such men as Mick Buck, Mick 'Spreader' Wright. Neil 'Nosher' Barnett, Mark Hunt, Neil Broster, Norman 'Granddad' Hammond, Neil Hammond, Glynn Pollard, Gary and Graham Ison along with back-up men Nigel Croft, Andy Corner, Paul Croft, Gary 'Spinky' Lewin, Neil 'Bun' Underwood, Tommy Wheeldon, Andy Tunks and more. All big hitters with that 'go getter' attitude. It made my heart glow with pride to see them go in the pit knowing that they would put in a good shift, it's what gave them a high, causing them to shout and laugh in the showers and canteen at the end of the shift. I mention lots of names because they all meant something to me for lots of reasons and they will become more clear later in this chapter.

Sue was more than happy with her school job and even took on cleaning jobs for a few elderly ladies nearby, as the kids grew more and more independent. Sue packed in her cleaning jobs as the old dears passed away eventually, one by one and she suffered from the grief she felt, becoming really closely attached to them. Andrea and Stuart were of drinking age now and didn't we know it!. Friday was the girls night, Andrea and her friends would call at the local supermarket on their way from work and invest in some booze, then go there separate ways and meet up again at our house, dolled up and ready for the evening, but not before they'd consumed the previously bought booze. By eight o'clock I'd had enough of loud music, shouting and hysterical laughter, all increasing in volume in direct proportion to the alcohol being consumed, that included Sue as she became one of 'the gang' for the night. I'd left and gone to the pub. The merriment still in full swing and then at nine o'clock they'd decide to go out for a night out in town.

Saturday night was boys' night, home from playing football Stuart would arrive and disappear upstairs shouting back to

Sue, "Are my shirt and trousers ironed and ready mum?" "Of course they are Stu, they're hung up in your bedroom as always on a Saturday." I often wondered what reply I would've got had I asked the same question on a Saturday night. I don't think it would be printable. At seven o'clock there would start appearing Mat 'Cat' Catton, Wayne Czervoski, Gav Baxter, Dave Haylock and Andy 'Tommy' Tucker. They would gather at our house before marching off as a group to town, with Sue saying," Don't they all look so handsome?" "Too young for you Sue," I'd say, disappearing through the door to the pub as her thrown slipper hit it and before any verbal abuse could be heard.

Sue wasn't so fond of them on Sunday morning, Tommy had asked to be awoken at 6am for work, which was why he was asked to sleep with his head near the door of Stuart's bedroom, the others scattered over the floor in sleeping bags and whatever, selflessly laid out by Sue after I'd disappeared to the pub for an hour. "I daren't go in there on a Sunday morning, it reeks of vodka and Red Bull god only knows what I'd find". She'd cringe. Tommy never did make it to work on a Sunday on time, for all his quoting of 'I'll sack you if you don't turn up on time'. From his boss. Anyway back to work for me at Asfordby.

Some unusual occurrences happened at Asfordby with pit men working underground and on the surface preparing for the first JCM's (Joy Continuous Miner) installation. One such operation was the erection of an arched tunnel to offer the JCM into for compatibility tests and it continued through the night shifts. It was sited well away from the main buildings of the time and was a good walk to see the guys and check on progress. Normally I would have taken the site Landrover but it was a lovely summer night and Eric Kupps and I had been discussing the merits of motorcycling and he had ridden his Honda CBR600R to work that evening. I was keen to try out his special machine and so coaxed him to let me ride around the perimeter road on it and he agreed somewhat tentatively. I arrived at the compatibility site seconds later and announced myself to a startled Terry Hughes, Mark Hunt and Nosher Barnett. "My god Don we knew that this was an ultra-modern pit but coming to see and check on us riding a new motor bike, surely not!" exclaimed Terry. "It's the way it's going to be now," I said. "And if I have my way we'll have bikes underground too." "We know you would, you're always racing from one place to another." They laughed, not knowing how close we actually did come to having push bikes underground. One to tell the grand kids about 'The night I went to see my men on a motor bike'. Eric was a bit concerned how long I'd been and feared the worst, but all turned out well.

Whilst on the subject of motor bikes, the deputy manager, Phil Marriott had talked of getting one and was importing an 883 Harley Davidson Sportster from USA, but it had not yet arrived. "I dress in my leathers in front of the bedroom mirror going 'vroom vroom' to get the feel for it," he laughed. Soon it came and on a fine Sunday summer morning he rode it to work. The business of the day was carried out and with his permission I took it around Melton Mowbray for a spin and bought it back unharmed. Harley's aren't my sort of bike and at 55mph top speed I was singularly unimpressed, but I didn't tell him that.

The wheel wash was a constant source of amusement in those early days, meant for lorries leaving site, it pressure washed their wheels and chassis and relieved them of excess mud. If it was overused although it recycled its own water it had to be topped up from the site supply as its own tank became depleted. A carload of sinkers decided on finishing their shift, would follow some lorries through the wheel wash and clean their car. Following the last lorry they drove to the marked position and the pressure jets operated for a short time and stopped, the car engine cut out, so the sinkers got out to push. Now refilled the wheel wash restarted, wetting everyone through so they moved away until the wheel wash stopped. They advanced on the car and pushed again, the motion setting off the sensor and pressure washing the sinkers again. They were soaked three times before they got the car free and running again.

Another time Andy Brown, a deputy, decided he'd take his motor bike through it. Andy, knowing how it worked, switched off the pumps and positioned his bike where he thought was the most advantageous place and started the pumps and watched as his bike was sprayed. The position not to his liking he waited until the spraying stopped and went to move it, this movement set off the sensors and soaked him. Dashing away his bike fell over, he waited until the sprays stopped again, thinking he had time to right the bike. This time the tank was replenishing and started as soon as the tank was full and soaked him again. No one witnessed this episode but a soaking wet Andy returned to the deputies cabin cursing the wheel wash with a mind of its own, drying himself off and using his clean work wear to ride home in.

'Born Lucky' Book Two: From Mynah to Miner and Beyond

NCB Landrover 1988 Asfordby Mine

Wheel wash

Every Friday afternoon Mr Skelding (Manager) and Mr Boyle (Chief Mining Engineer) would like to walk around the site to see the places they didn't normally see and would want me to accompany them, weather permitting. On one such occasion we were about half a mile from the shafts in the washed coal storage area that was to be, just chatting as we walked about nothing in particular. When Mr Boyle spotted a guy some 200m away doing a spot of brick laying to a manhole cover. "He hasn't got a hat on, Don. Go and tell him to put his hard hat on," he instructed me. I glanced at Mr Skelding, he glanced back as if to say 'humour him'. I trudged across what would pass as a ploughed field wondering how I would explain to a man essentially in the middle of a field why he should wear a hard hat. On reaching him I could see that it was a friend of mine from Melton, Joe Maguire, "Joe, don't ask any questions, but put your hard hat on." "Righto Don." And immediately put it on. Next time I saw him in the pub he had told the whole darts team. "Did you think the sky was going to fall on his head Don?" I had to endure that all night long.

Concreting ops in the shaft

'Born Lucky' Book Two: From Mynah to Miner and Beyond

Downcast tower under construction

Upcast tower under construction

Whilst the shaft was stood down for the freeze to take place, the permanent concrete tower was built around the sinking head gear and before sinking could resume. The emergency winder had to be bought on site and a dry run had to be done to remove any snags there may be. The winder site had been identified and a concrete base laid with exact markings made as to where the winder should be parked. It was again on a Sunday morning, full rescue team in attendance, British Coal staff headed by Ian Hollick (Safety Officer) and deputies and I'm not sure if Lynne McDermott wasn't there, she was the 'constant' at Asfordby and kept us all in line, she was the managers PA.

The plan being that Tony Spring and I would go up inside the tower, locate the appropriate rope slot, coil a long rope and toss it through the slot to the waiting crowd below next to the winder. They would then attach our rope to the winder rope and we would pull it up and thread it around the relevant pulleys. With everyone in position Tony and I had a couple of practices at getting a looped 50m of tow rope through a 15cm wide slot in the concrete. Now fully prepared we whirled the rope like a lariat and straight through the slot and into the clear air outside the tower. Unfortunately the wind caught it and instead of it coming to rest at the waiting crowd it was blown around the tower and entangled itself in the scaffolding still in place, we couldn't pull it back, it was fast, so they had to scale the scaffold to free it. Everything went without a hitch after that. Best laid plans and all that............................!.

As I said earlier the inseam boys were getting established with their five-man crews and setting arches and bolting the roof as well. The pit bottom crews of eight men per shift had a very steep learning curve to negotiate. Not only had they never seen anything of these proportions, (7.5m x 7.5m) and steel the size of which couldn't be manhandled (25cm x 20cm) but they had to learn to used Holman rotary percussive drill machines coupled with air legs. They had to get confident with taking a Dosco (40 tonne) down and up a 2.4m ledge but the steel they set had to be within 2mm of tolerance going around a bend and fully grouted because it had to last 50 years, the planned life of the mine. The same went for the nine-man teams in the Markham/Robbins Shield headings.

Roger Cornish, Russ Millar, Phil Forrester, Kev Hines, Chris Bircher, Chris Wilson, Graham 'Kermit' Hall, Bert Colley, Steve Hovell and a host more that made a name for me and weren't overawed by anything I placed in front of them.

We were put onto six day working, Monday to Saturday, days, afternoons and nights and then nine days off. Some lapped it up, some didn't mind it, and some never did take to it. But it was agreed, and we had to make the most of it. The downside of it for me was twofold, I didn't get a day off on my week off as I was needed by my boss Terry Pennick, ah well I was getting paid well for it. The second thing being that the maintenance period was only twenty-four hours as opposed to 48 hours and the stuff we had to attempt and complete to avoid delays in production was nothing short of phenomenal at times. The pressure was always at maximum on Sunday afternoon shift as production started Sunday nights. The get things completed on time I had to have some incentive for the men, I couldn't pay them more, mid-week overtime was a no no and weekend overtime was at a premium and they were already working it. I reverted back to the only thing I knew, job and knock.

I only applied it Sunday afternoons to ensure production started on time. I would be in the control room at the end of Sunday afternoon shift and one of the time office lads, Darryl McDermott, Stuart Farmer, Adrian or Kev would come to me and ask, "How shall I pay them?" "A full shift," I always answered. I thought I had that power, and the time office lads did or else they wouldn't have asked, it was always a custom and practice as far as I was concerned.

Unbeknown to me the other three overmen were doing the same and I don't know how management found out, but they conducted an undercover investigation with those interviewed sworn to secrecy, we as overmen weren't aware of this investigation. We were called to the mine one morning and gathered in our office and one at a time were sent for by the manager Mr David Betts, Bill first, me, Frank and Nigel. Each in turn answering questions. One I strongly objected to was 'Have you instructed banksmen to falsify a statutory report'. And I told him so, I also said that I was willing to take a polygraph (Lie detector) test to prove it. We were all four of us suspended with immediate effect. I cannot describe how that felt, I was numb to the point that it was funny, ridiculous, unreal but sickening. Frank and I went to the pub.

That suspension lasted a week, the mine did next to nothing, and deputies would only report on matters of safety and nothing else. To a man the workforce respected and backed us, as I've mentioned previously how much I respected these guys back it was for that reason, they cared about how we'd been treated. I sometimes think that management had been looking for a way of taking our respect and power from us overmen, it didn't work. It was a fortnight before I started back, the suspension ended the day I was leaving for the Isle of Man TT races and no matter how much the manager demanded I come to see him on the Saturday, I refused. I wasn't missing the TT races for anybody!. And f**k 'em all, that's how much I was hurt.

Spontaneous combustion can be prevalent in some coal mines depending upon the volatility of the seams being worked. It occurs when air gets into contact with exposed coal surfaces and a slow but consistent laminar flow is maintained either through wastes, known as 'gobs' behind longwall faces or along roadway edges and corners. It is the reason the mine spoil heaps are seen to catch fire at times. Asfordby was particularly prone to 'spont Com' as we termed it, and regular inspection of the roadsides was carried out. It could be alarming to view a perfectly normal roadway side inseam with the naked eye and then view the same roadway side through a heat seeking thermal camera. It could be seen to glow red along spillage areas and corners even though it may seem normal to the touch.

Mine air monitoring played a big part in the detection of 'hot spots' within the mine and teams of men would be deployed immediately to a suspect area, even if it meant stopping a machine to deal with it. I've been drilling myself holes for injection of inert solutions to quell areas and on withdrawal of the drill rod, flames have shot out of the hole. Of course that hole was immediately injected. It was an ongoing battle and John 'Miffy' Smith had an important role at Asfordby, deploying his teams and supplying us overmen with numerous plans of the many junctions affected and being worked on.

It was a sad, sad day for me when I left Asfordby, I had been a part of something special. I had built some tremendous constructions normally done by contractors and above all I had the privilege to have known incredible men and deputies.

All New

At 47 years old, I had made it the house paid for and now retired with enough money to take me through to my work pension at fifty years old. Sue wasn't as keen as me on the idea, but I spent the next thirteen weeks gardening, decorating and holidaying with Sue. The kids had met their respective partners and were getting along just fine and all in the garden was rosy. But there's only so much of that you can do before you start the cycle again of re-decorating etc. I took a job at a factory, it was fine at first but I suffered a reaction to my skin from the powdered resin and fibre glass involved in the processes and so it came to an end. I went back to the mines and took a challenge at Daw Mill in Warwickshire initially for five weeks but ended up being for five years. I was in a tunnel there for three years or more, I loved these big circular tunnels, must've done, since leaving Ellistown I'd done very little else!.

That part of my working life ended and I'd eligible for my works pension for two years and not taken it. Now was the time to start it I thought, I'd worked five years longer since my last shot at retirement, maybe I'd be more successful this time. Not a chance, I lasted ten weeks this time before I was climbing the walls and driving Sue up them too. Along to the job centre

I went and saw an advert for a valeter at a local Vauxhall dealership, Aston's. I applied and went to see DJ (Oldham) as he was called, an older man, very informative, having a pilot's licence and health conscious.

As co-owner DJ took me on a tour of the site and offered me the job, I told him that I didn't want full time employment, but I'd give it a try. That was Monday and I was to start the following Monday. DJ rang me at home on Wednesday and said that he'd read my CV and didn't think valeting would be fulfilling enough for me and could I come to see him the following day. He explained that they were looking to set on someone that would collect and deliver cars all over the country, driving one way and using public transport the other. My eyes lit up, I loved driving and hadn't been on a train for years and years. Needless to say I accepted.

My first trip was organised by a young man called Peter Bettelley, a shy lad but very conscientious and keen to do well. It was to collect a new car from Hayward's Heath. It meant catching a train to Leicester, on to London St Pancras, on to the underground and around the Northern line to Victoria station, getting a train to Hayward's Heath and ending with a call to the local Vauxhall garage to come and pick me up. It worked a treat, and I thoroughly enjoyed it, no sat navs in those days but once on the M25 I was away and dealerships were only too pleased to help tell you the shortest route. I learned very quickly from Brian Kaye, cCo-owner at the time) a very astute and efficient businessman, as how to examine a vehicle and record it, how to describe it over the phone, how to negotiate and to trust no-one within the car industry. It was a more responsible job than first meets the eye and a lot of money can be lost through bringing a bad car home or even a stolen one, hence all the checks and responsibility.

I travelled all over the country with Pete's help and organisation, we became good friends. Another source of incoming cars was ex-rental models, all Vauxhall of course. This under the jurisdiction at the time of Mandy Stevens, Mandy was great, tall, great sense of humour with wild bushy hair. I think it's fair to say that she was a bit afraid to give me jobs initially, don't know why, maybe thought it was a bit too much for me, but it wasn't, when you enjoy something it's not hard work. Two more sources of work were returning serviced vehicles for Mick Knight (Service Manager) a brilliant, very active and genuine guy, to the local RAF and army stations of Melton, Cottesmore, Witterinng and Edith Weston. Also for DJ and Stuart Smith (Body shop Manager), Stuart was and still is a keen darter, he never let me forget when he bought a team from Shepshed, challenged my Boat team and beat us.

After a few months and DJ and Brian were sure I wasn't a fly by night, Brian asked 'would I be prepared to drive a small car transporter with a beaver tail' (a lowered rear section as part of a ramp). The trains had been awful of late due to inclement weather and a massive programme of track replacement nationwide due to a rail disaster in Clapham. I answered yes. It was another few months before I collected it from somewhere in Norfolk I remember.

With the Ford Transit 2500cc transporter on site I was tasked with setting it up with straps to secure the vehicles to it, the on-board loops for attachment of them already in existence. Four sets of ratchet straps later and four loops made of the same webbing and stitched by a shoe machinist saw me ready for the road. DJ asked me to get a car from the lot and load it on the vehicle as a trial and I did it, Brian and he came to inspect it and were delighted. Apparently they'd had one before and the loaded vehicle had come adrift, fortunately with no consequences, the truck was stolen shortly afterwards. I spent many hours in that truck and a part-time job very soon became full time. Following are some memorable experiences and adventures in it.

A Day with Tom

Tom was a new lad that I hadn't met yet, Brian was picking him up on his way to work arriving at 08:30am I assumed. I always planned the day before and made any necessary arrangements and paperwork. Knowing Tom was a young lad with a driving licence I guessed his knowledge of the country would be sketchy and so I had printed off the necessary maps and even wrote out information as to what junction numbers and major roads etc.

Brian arrived with Tom and introduced us and I started to explain the day's work, he told me he was straight from university, when DJ interrupted and led Tom away for a tour of the premises and induction. The plan was reasonable for a Friday with no delays, we were to take a Vauxhall Astra and deliver it to Chester, make our way to Manchester and pick up another Astra which Tom was to drive back to Aston's whilst I carried on to Accrington to collect a Vauxhall Corsa, busy day but doable. DJ had finished Tom's induction and cost me, in my eyes, almost an hour that I hadn't accounted for.

I explained to Tom what the day entailed whilst driving and he said that no one had told him that he would be moving cars around the country on his first day. I could tell that he was uneasy by the way he kept shuffling the maps and write ups that I'd given him. I did my best to reassure him and that really it was simple but for someone on their first day nothing is simple if you've had no grounding in anything. We made our way and found the address in Chester, unloaded the Astra for the customer to inspect and then went into their house for payment. They were a couple that had just arrived from South Africa and were very pleasant offering us tea whilst we waited for the card payment to clear. It didn't clear and so we waited and waited until it cleared. I had lost another hour and I wondered now if we'd make it home before the dealership closed at 6pm.

Arriving in Manchester I inspected the second Astra and payment was made. I put the trade plates on it for Tom to drive back to Aston's, I fuelled the car for him and set him on his way, going through with him again the route back to Melton Mowbray. I set off for Accrington, it was chucking it down when I got there, I got soaked loading the new Corsa but at least now I would be straight onto the M65 heading for the M6, with Tom well on his way by now. Better ring him to see what progress he'd made. "Where are you Tom?" I asked "Don't know Don." "Tell me the last junction you passed if you're still on the M6." "Don't know but I'm approaching fourteen now." "Get off at fourteen and turn around and come back to fifteen A, that's the A50 to Kegworth." "OK," he answered. "Let me know when you're on the A50 Tom." "Will do."

"I'm on the A50 Don," Tom informed me. "Good, keep on that until you get to the M1 at Kegworth and then take the A6 to Loughborough, no need to remember it, it's marked on your map and written down." I was bombing along the M6 southbound and going over the Thelwall Viaduct and making good progress for a Friday afternoon. Time was flying and so was I and when I reached junction 15A I thought to check on Tom. "Where are you now Tom?" "I think I've taken a wrong turn again Don, I'm on the M42." "What junction?" "Don't know." "Get off at the next junction, ring me and keep on driving around the island telling me any signposts that you see Tom." "OK Don, I'm not very good at this am I?" "The best thing to do is drive slowly so that you can read the signs and not panic," I advised Tom.

He described the Flagstaff Island, I told him which turn off to take and kept him on the line until he was on the Rempstone road, that would keep him happy for half an hour. I was now on the A50 leaving Stoke and heading for Uttoxeter. "Tom, when you get to Hathern you have to cross the A6, do a left and immediately do a right." "I've done that Don and I'm in a pub car park called the Brewers Fayre." "You've done a right and left and the pub's called the King's Arms," I said to him. "Double back and do it again and tell me when you get to the Rose and Crown." "OK," Tom affirmed. At that moment I had an awful thought about the fuel in Tom's car, I'd fuelled it as if I'd driven it with no missed turns and Tom had had a number of detours. "I'm at the Rose and Crown now Don," was the update from Tom. "What fuel do you have left Tom?" "Not much." "Is the fuel light on yet?" I asked "Yes." "Is it flashing?" "Yes." "Turn around and come back to the fuel station past the Kings Arms, you haven't enough fuel to get you home." I was now at Kegworth. "Pull up at a pump and put in £5 worth of fuel." "I haven't any money," panicked Tom. "No sweat, the attendant won't know that, and I'll be with you with my fuel card in two minutes," I assured him.

Tom refuelled and said, "I've never been so glad to see anybody in my life. How the hell did you visualise where I was all the time?" "Luckily I know that part of the world very well. Follow me now all the way home to Melton." We landed at Melton at seven o'clock that night, Mick Knight had stayed on to secure the vehicles and Brian had stayed on to get Tom home. Tom was with us for a while and did other trips with me, none so eventful as his first, thank goodness!.

Flying High with Max

Max was the dealer transfer man, arranging the transfer of cars/vans between dealers sometimes new, sometimes not. He was a younger type of guy, not inexperienced, tall and slim. There were two cars to collect in Scotland, one in Kilmarnock and one in Aberdeen. Mandy Stevens arranged it and Max had done the paperwork. The truck was in the service bay having a new engine fitted, I was to have a total of four engines fitted to that truck, even with regular servicing by our own technicians, and so wasn't available.

The plan being that Max and I would make our way to the dealership for five thirty in the morning, there Mandy would pick us up, as she did in her nightie and dressing gown. She then drove us to East Midlands Airport to board the seven o'clock Glasgow flight with our trade plates and paperwork. The flight was an hour and knowing the task ahead I knew that we hadn't time for local transport from the airport to Kilmarnock and so I hired a taxi. Arriving at Kilmarnock after the thirty-mile trip we identified the car, it was as described, paid for it, phoned in to Mandy whom was now at work, and carried on driving from Kilmarnock to Aberdeen where our second car was located.

The hundred-mile trip taking just under three hours and were at the dealership for the second car, going well I thought. On viewing the car, a Corsa, I was less than impressed. It was obvious that it had been a young lad's and had been customised in a less than professional way, laid back seats, stainless steel rally foot pedals and other stuff. I complained to the dealer, "I don't think you understand, we rejected this car from you, and you are collecting it at your cost." "I beg your pardon, you're right I didn't understand," he said. "Can I have the paperwork from you then please?" I asked. The paperwork was minus the logbook and a discussion ensued.

"Hello Mandy, can you have a look in the file for the logbook to this car, they are saying that they've never had it." "I will do Don, I'll do it for you, I'll ring you back." She was always soft soaping me, in case she needed a favour in the future.

Time was passing and we were a long way from home. Mandy rang back and said that there was no trace of it in the file, "Max should have sorted this Don, have a word with him," Mandy said quietly. I asked Max about the paperwork and the logbook, he searched through his briefcase and having removed old sandwich wrappers, a flask, and empty drinks cans he finally got to the temporary file, opened it and there was the green tear off slip of a logbook, pertaining to the vehicle that suffices as ID for the vehicle. I was not best pleased, and Max knew it. That wasn't the end of the delay, the car still had outstanding finance on it, no dealer will accept a car with outstanding finance on it. I was beginning to understand why Brian had sent Max with me rather than Chris Thurtle, my regular mate, I did a fair few thousand miles with Chris, but more of that later.

Another call to Mandy to help me out, she sounded apologetic, but I told her that it wasn't her fault, again she whispered, "This was Max's job to sort out." Mandy pulled out all the stops to get the finance paid off. I was fuming by the time that the document of 'No further interest' came through by fax from the finance company. "I'll drive the

Corsa." offered Max, knowing that it was a shed of a car. "Too right you f**king will. Check it for fuel, oil and water," I seethed, we'd just lost three and a half hours for the sake of ten minutes work at a desk that should've been done yesterday. Max had to climb in the Corsa through its boot, both doors were jammed and he couldn't work the rally foot pedals with his shoes on. That made me feel better. The journey home took eight hours, and I landed home at midnight. Max wasn't with us long.

With Chris at a Push

Nice day's work in front of Chris Thurtle and I, Chris was an ex police officer, was a pleasant able man and very helpful, reliable and organised. The task ahead being, deliver a vehicle to Tunbridge Wells, below the M25 on the truck, carry on around the M25 to Kempton Park area and pick up two more, Chris to drive one back and I with the other on the truck, simples. Almost a full lap of the M25.

Turning off the M25 and heading for the dealership I stopped at the side of the road to get the last mile of directions from our destination, no sat navs in those days. I was all ready to set off again, but the engine wouldn't start, just the sound of the solenoid clicking. "I'll ring Mick for advice, you unload the car and take it on trade plates to the dealer and ask one of the salesmen to come back with you with a battery booster pack and see if that will help," I told Chris. That happened and Mick mentioned one or two things to try, to no avail. The booster pack didn't help as it was flat too!. I said to Chris "I wonder if the starter motor is jammed, let's put it in first gear and rock it back and fore to free it". We did that with no result. "We'll push it and I'll slam it in gear to free it." So we did and the engine fired but died. "We'll push it back to the bus stop (5M) and then we'll push and I'll jump in and bump start it," I said enthusiastically. "Push a truck, you must be mad," said Chris, rather less enthusiastically. One push and she fired, I narrowly avoided a tree but all was well again. "What now?" said Chris, "Do we head for home or carry on?" He already knew the answer to that.

"I'll drive to Kempton," Chris volunteered. We were mobile again, I phoned Mick and Brian to let them know. "Don't switch the engine off when we get there Chris," I instructed. We were a half mile from our destination when an accident happened in front of us with a car ending on its side. The police were soon in attendance and let us through, and immediately we were past, the garage was there on our right, Chris turned in and parked, instantly turning the key and the engine went silent. "We'll get them to give us a push," I said convincingly. We identified the cars, verified them, had the paperwork and paid for them but they were across the road in a secure compound. "Do your stuff Don and get them to give us a push," Chris said, smirking, he didn't have my confidence and didn't think they'd help, these southerners.

I went to the workshop and returned with four guys. The forecourt was very small and was riddled with potholes, making it difficult to get any sort of a run, would that bl**dy truck fire, but just as we were about to call it a day, away she went. "Take it over the road Chris and DON'T SWITCH IT OFF!" I emphasized. He did that and we loaded one car and Chris followed me home with the other. It turned out that the alternator was defunct and so was its warning light bulb on the display.

Elastic Band

I was out and about with the transit transporter, just delivering and collecting sales cars locally. I happened to be passing through Langham, a village just outside of Oakham the capital of Rutland County. Going uphill I felt a loss of power and looking in my mirror, thick black smoke was enveloping the following cars. Luckily there was a lay-by ahead and I pulled into it, raising the truck's bonnet everything looked in order, but it obviously wasn't. My first thought was to get the car back so I arranged for two young lads to come and help, Tim Moore and Marc Burton, one to drive the other to me and then the passenger would drive the second car on trade plates back to the dealership.

I am not one to lay back and wait for something to happen and so I gazed at the engine again and tapped, knocked and pulled things. To my surprise a rod, about as thick as a drinking straw came loose in my hand, one end being a ball joint and the other a hinge, the ball joint was worn and had jolted free. Reckoning it wouldn't take much to secure and hold it in place, I hunted around the cab for something suitable, String was no good, inflexible, I needed something with a modicum of 'give' in it. I have a habit of picking up loose elastic bands and there was one, just the right thing that I was looking for. I then proceeded to bodge a temporary joint to get me the nine miles to home. Pit 'get you by' skills never leave you. One to tell the grand kids about 'The day I drove a truck home on an elastic band'.

Bonny Scotland x 3

The truck had had another engine change the day before and I was apprehensive about taking her out on a long trip, first outing so to speak. As per usual I loaded the car the day before and took the truck home for an early start to Glasgow. At 5:30am I set off, much to my wife Sue's annoyance, I'm not the quietist about the house of a morning. Clearing Melton and heading for Grantham with very little on the roads. Every trucker will tell you it is the best time to travel, through the night. The plan being to head up the A1 to Scotch Corner across the A66 to Penrith and off up the M6/M74 to Glasgow. Great.

Having got on the A1 and climbed the hill I was now going flat out downhill towards Great Gonerby and the services. Oh no, I was losing power, the truck was just coasting and so I pulled into a lay-by, opened the bonnet and could see a gap where the glass diesel filter should have been. Nothing more to do than walk back up the hill to see if I could find it hopefully intact. Trudging uphill on a grass verge at 6am on a winters dark morning makes you realise just how frightening the motorways are. That wind rush as vehicles pass is horrendous. I carried on to where I thought it would be, but nothing and so I about turned and walked back to the truck, too early to ring Mick or the customer whom may not want his new car driving to Scotland.

I made it to the truck and shone my torch around the engine bay and to my surprise, nestling below on a heater hose was the missing filter bowl complete with sealing ring. I quickly replaced it as tight as I could by hand and turned the engine over and she fired eventually and yours truly was on his way again with no more problems that day. I did let Mick know and he advised calling at a service station and have it inspected and tightened securely.

Same thing again, taking a car up to Glasgow via the usual route, A66, loved driving that road from Scotch Corner to Penrith. Hammering away as usual when there was an almighty bang and thick black smoke obscuring the cars behind. Again, luckily there was a lay-by and I managed to coast to the far end and onto the grass verge. Opening the bonnet I could see a gaping hole in the side of the engine block, she was dead, bless her, confirmed by the oil slick and the piece of cast metal on the road behind me.

It was 8:30am and Mick would be at work, I phoned to let him know and he said he'd arrange pick-up of the truck, I told him it was on the A66 at Appleby. With that I called the customer and gave him the option of having his car today as arranged but driven by me or having it transported, but I didn't know when that would be. Of course he took the former. I unloaded the car, put the truck back together, wrote contact details and put them in the windscreen and locked the truck. I then put the keys to the truck on the front wheel (standard practice with a locked vehicle in the trade) and continued on my way.

Mick rang half hour later and asked where I was. "I'm at Carlisle now." "WTF you said you were broken down! I've arranged pick-up and you're bombing up to Glasgow." He wasn't pleased but he calmed down when I explained to him. "Just wanted to confirm where the truck is mate," I told him and he ordered another engine. The customer expected a major delay in delivery, but no, I explained that I would be with him before the arranged time as I would be driving a faster vehicle. I then drove his part exchange Mini home to Melton Mowbray. The trip went smoothly after that.

The most rewarding thing about my job was delivering new cars, the joy on the customer's face on seeing their new car arrive coupled with the care, attention and ability after a long journey by their delivery driver. One such trip was to Rose Hearty, just along the coast from Aberdeen. (What a lovely name for a village). Brian had this business acumen where he was able to sell new cars at least a £1000 less than other dealers and still make money. I rang the lady of the imminent delivery of her new car, it was something I did when I was around an hour away. It worked two ways, it was an appreciated courtesy call from me and it told me that there was someone available to receive the car and that they weren't out shopping.

I arrived at the road, an avenue of trees lining it each side, I proceeded slowly in the fading daylight looking for house numbers. I needn't have bothered, the lady was in the middle of the road waving vigorously. I pulled up in front of her and got out of the cab, she came up to me, put her arms around me crying and sobbing loudly. "You've no idea how relieved I am to see you," she said in her Scottish accent through her tears of relief. "I did call you yesterday and again an hour ago from Aberdeen," I said soothingly. She and her husband had thought the deal too good to be true, they had parted with thirty grand of their money for the new Vauxhall Insignia and were yet to see it in the flesh, so to speak. They had thought it may be a scam. I now understood the relief of that lady, she immediately called her husband aboard his vessel (he was a ship's captain) in the English Channel to tell him the good news. I'd imagine he'd splice the mainbrace that night.

The day the roof caved in

Easy Peasy job this day, aren't they always the ones?. Just mosey on up to Ellesmere Port near Chester and deliver a Vauxhall Insignia VXR8, a beast of a car and beautiful with it. I had the two-car transporter that day and I'd raised the top section and loaded the car on the lower section, lovely. Heading for the M6 along the A50 for Stoke looking forward to seeing the Vauxhall plant that made Astra's. I reached there in good time, found the plant and got through security, blimey this place was mahoosive. Driving slowly on deserted site roads, each little island or turn off had a massive signpost with about fifteen to eighteen destinations on it with little arrows giving the directions of each destination, I was looking for gate 33. I was getting more confident just slowing as I approached the signs, I'd a feeling I was getting close and I was straining to read this sign when there was an almighty bump that shook me on my seat. I thought nothing of it and on seeing my gate number I turned in saw the exchange car and commenced unloading the VXR8. I had called the customer and he was on his way, and he brought the keys to the exchange car and I was loading it, when he came to me and said. "Don I have a problem with the car." "What's that Ray?" I asked. "Look at the roof." I was standing on the lower deck of the transporter and could see the roof from above. I've never felt so empty in my life, there was a dint right across the roof, maybe 10cm deep and above where the passenger and driver would sit.

I couldn't apologise enough but Ray was calm and said, "After what I see in there (The factory) on a daily basis, nothing surprises me." I rang Brian and he advised me what to do and assured Ray that he'd get another one for him before the end of the month. I unloaded the exchange car and reloaded the VXR8 and I could see that the dint lined up exactly with the cross beam of the upper deck. Then suddenly

I realised what had happened, that bump before the island whilst I was trying to read that sign!. Driving very slowly on the way back through that area, not only was there a big pothole but exactly the wheelbase length of my vehicle away from it was a low speed bump that had magnified the effect of the pothole.

Brian accepted what I said and could see how upset I was, as I always was at any damaged vehicle. He didn't berate me but asked me to go back and photograph the area and pothole. I even stood my clipboard in the pothole to give perspective to it, it almost disappeared in that crater. I didn't hear of the incident again, I hope a claim was made and accepted by Aston's. I didn't feel good about having nearly written off a £35 grand car!

Time's Up

It must've been around 2010 Dave Hebb came to me in the Mash Tub in Melton, "Do you play golf Don?" he asked.

"Not for forty odd years, and I didn't break 120 then," I quipped. "I've got some invites going spare for an event to do with the Holwell foundry works, it's very friendly and it doesn't matter about ability. It's twenty quid and includes a bacon cob and coffee, a round of golf and a three-course meal," Dave informed me. I rather liked the idea and went for it along with Ron, Alun Davies, Clive and Steve Rodrigues. 'Fail to prepare, prepare to fail'. I've always believed that and so I borrowed my son-in-law Carl's golf clubs and headed for the driving range. I surprised myself, I wasn't at all stiff after hitting a couple of hundred balls and had remembered some of the basics that 'The golf doctor' John Jacobs had taught all those years ago.

"We'd better have a practice round before the day," Dave suggested. We all thought it was a great idea and so it was arranged. I went around the Six Hills golf course in 106 that day. To say that I was astonished would be an understatement and I didn't get near that score again for a while, but the golf bug had well and truly bitten me. It was the start of a major part of my life for the next few years.

I had wondered again about retirement, I was now 62, still with a zest for life. I had reasoned that I couldn't go from working full time to doing nothing after finishing work and so I approached Brian about going part-time, his answer took me aback. "Don, go away and over the weekend think about what you'd like to do and let me know on Monday what your answer is. You can finish, work part-time, one, two, three or four days a week. You can deliver parts, do site maintenance, whatever you want just tell me and you can do it." I couldn't have wished for a better answer than that.

I chose to deliver parts much to the delight of Sally Oldham (DJ's wife) she'd wanted to work part-time for a while now, bless she was a lovely lady. I chose to work Tuesday and Thursday, Brian was as good as his word, and it suited my plan of playing golf on Fridays with the boys. That carried on for a few months, it was nearing Christmas and my replacements in sales delivery hadn't been successful, Brian let it be known via Mandy that he'd like me back in that department. I knew what it would be, it would start part-time and grow to more than that and I told Mandy that I didn't want it and that I was considering retirement. "Oh don't finish Don," she said. I told Sue of my decision, she said, "As long as you keep from under my feet."

At the end of November I took an envelope marked for Brian and Mandy saw me with it as she parked her car. "Is that what I think it is?" she asked. "It is." I looked down and answered. "I'm not opening it then," she said laughing. It was my notice, and it signified the end of a very challenging but immensely rewarding and enjoyable ten years of my life.

Golf filled the void and on the days that Sue didn't need me I took myself off to the range to practice for the upcoming Friday's golf that became a competition of pairs, alternating partners each week. One such week there was only three of us and we decided that we'd play at Rutland Water Golf Club. It was raining heavily, and Dave and I had picked Ron up at home in my car. Discussing the weather Ron quoted, "Rain at seven, fine by eleven, is what the farmers say." With that, off we went.

Waiting in the club's car park, it still hadn't stopped. "I think it's abating," Dave uttered. "Let's go then," Ron suggested. So bravely we strode to the Pro's office and paid our dues and proudly marched to the first tee and practised our finest swings before teeing off in the rain. That round of golf was a nightmare, we didn't see anyone, we hadn't proper wet gear and we were frozen and wet to the skin. Hebby said "F**king farmers, what do they know?" Another similar occurrence was wintertime and Dave and I played. "We must be mad, it's minus five this morning." Dave complained "It'll warm up," I said hopefully. Frost was on the fairways and the ball bounced around like a pinball, the greens were also frosted over, and the golf ball was twice its size with ice before it reached the hole. We cut the round short by three holes and got back to the car. "It's still minus three," said Dave. "Told you it would warm up," I said, blowing my hands and rubbing them vigorously.

During my time at Aston's Stuart had got married to Maria and moved into a house with her, with that I inherited a wonderful grandson and granddaughter, Jake and Leanne. Andrea had met Carl and they were modifying a house together and had taken Casey her dog with her, and later presenting us with two beautiful granddaughters. The house seemed empty and it was, something we'd have to get used to but I suggested a motoring trip abroad taking in France, Luxembourg, Holland and Belgium in the Jag. Sue was hesitant, she wasn't much for foreigners but she said yes and to be fair she really enjoyed it after I told her we wouldn't be taking the ferry. She made it plain to me that she didn't understand why it was her that should pay all the tolls and car park tickets, she said that I was "doing it on purpose". I laughed and explained that I had a right hand drive car in a left hand drive country. Still miffed, she made it known that she'd get her revenge somehow and I knew she'd make me pay for it.

'Born Lucky' Book Two: From Mynah to Miner and Beyond

Stuart and Mariah

Andrea and Carl with daughters Ruby and Megan

After that we stayed in Britain and travelled a great deal at weekends and holiday times now I was in love with driving. "Let's go Jagging," she'd say, meaning she'd fancy a trip out, maybe to the Lakes, the Peaks, Snake Pass, or wherever the fancy took us.

I visited my dad in Ibstock every Saturday after mum passed, he was too old now to drive, Sue would come along, and we'd enjoy a fish lunch together from the local chippy. In 2012, I was getting ready to leave for the Saturday trip and Sue said, "You go Don, I'm not feeling too good and I'm tired." "OK," I said, "but you should see a doctor Sue." "I know." The next week was the same. The following week I was a little awkward with her. "You must see a doctor Sue, it's dragging on now. It may only mean adjusting your medication." "I know, I'll make an appointment on Monday morning," she answered.

The summer had been a wet one to that point, but the last three days had been warm as it should be in July. Saturday night was as normal, I'd been out from eight until nine whilst Sue watched TV and drank a glass of wine. "I'm cold Don," Sue said as we climbed into bed. "It's not cold tonight Sue, come here and let's cuddle, I'll keep you warm." She fell asleep but had a restless night, I could tell. At quarter to six next morning she awoke and sat on the side of the bed. I thought, 'she'll be going to the bathroom in a minute'. Sue didn't she fell backwards into the bed and onto me. "Sue, Sue," I shouted whilst running around the bed to her. I gently slapped her face with no reaction, shouting her name. I knew then that the emergency services were needed, she was looking at me with stationary eyes as if to say, ' Help me Don, help me please.' I quickly got her off the bed onto the floor and whilst checking for a pulse and her breathing I dialled 999 from the phone at Sue's bedside.

The operator asked the questions and advised me what to do, but I was already doing it, chest compressions and two breaths. My thirty odd years of first aiding in the mines kicked in and I was so grateful, there was no panic, Sue was not turning blue, and her chest was rising with my respirations. However long it was it flew by, still hearing the call handler and suddenly a knock on the front door, it was an ambulance. I ran down the stairs and let them in and showed them to our bedroom. Then another knock, a first responder, the ambulance came from Melton ambulance station when we had one and had beaten him. It was now 6am, what seemed like an eternity was just fifteen minutes, I remember that I was on my third two-minute cycle when the knock at the door came. It meant the ambulance was at my home on around six minutes.

They asked me to go downstairs, and a paramedic made me tea, they continued my work and I was sat just hoping but when a paramedic appeared on the stairs I knew Sue had gone. I've never known such emptiness, numbness with no thoughts at all in my mind for a few moments. The female paramedics were very understanding and did their job well, reassuring and comforting me.

I gathered my thoughts and wits and wondered what the hell I was going to say to Andrea and Stuart. I won't go into the calls but Andrea screamed out loud and Stuart's response was, "I'm coming round." Not calmly but in control of his emotions. Andrea quickly called Carl who'd just arrived at work to come back and look after the granddaughters that they'd blessed us with.

Maria came with Stuart and Andrea quickly followed. Of course they wanted to know what had happened, the paramedics were still here and I couldn't manage two sentences together without breaking down, no matter how many deep breaths and self-control I tried to summon up. We sat, drank tea and talked and talked, I was wondering whom to call next of Sue's many friends in Ashby and Melton. Each call meant having to repeat the events of the night and there were no two-minute calls, all were twenty and thirty minutes as the news slowly sunk in to whomever I had called. The undertakers turned up to removed Sue to a place of rest and just after I received a call from an old friend of Sue's, Mary (Tissy) Smith, "Oh my god Don what's happened?" I told her and she said, "You know what I'm going to ask now, can I bury her for you Don?" It was a bolt from the blue, thinking very quickly I realised Tissy had a daughter that had taken over a local funeral directors. "Oooh Sue would love that Mary, I can't think of anyone better." So that was settled.

Things moved on and as much as I wanted the world to stop turning, it didn't. All of the things that have to be done all take time and with the help from the family the day of the funeral came, and it went as well as these things can do. Old and not so old friends turned up, Tom Fothergill travelling from Northumberland to pay his respects. Someone asked, "It must be a burden to you, having to perform CPR on Sue and losing her." I answered, "Not at all, it's a comfort really. I did what I was trained to do, and I didn't make a mistake even though I've replayed it a lot of times in my head, trying to blame myself. No it's not a burden, it's a comfort."

I had to go through probate for some firms that would not accept the death certificate as the only proof of death, and I decided to do it myself. I would get up in the morning and look at the forms and then look at the other stuff that needed dealing with and decide to play golf, it didn't make things go away but it made me feel better, but alas they were there again next morning. Things became so bad that I rang my aunty Sue in North Yorkshire and asked if she could deal with a lodger for a week. "Of course nephy come up, I've got plenty of room for you." (She owned a B and B) So I went up to visit her and Ian her husband. I was running away really from self-imposed pressure, but it was a very welcome break and she put up with my late-night outpouring of emotion.

I managed probate after that and all was getting better, but I was still missing something. I bumped into Keith Tomblin a good and dear friend "Keith how much is it to become a member at Melton Golf Club?" "Come and pick me up and we'll go along and find out," he answered, he was at work with his printing business. So off we went, and I was made a member of the club. Although I didn't intend it, it took over my life and I was to be playing up to six times a week. I thrived on the competition and the friendliness of the guys at the club. So much was the time I spent out of the house Andrea and Stuart complained, "We'll soon have to make an appointment to see you." I was still running away I think.

Peter Marsden and his wife Tilly had a trip planned to Switzerland in September. "Would you like to come along with us if Chris and Katrin have a room spare?" "Try and stop me," I answered eagerly. I didn't say no to anything in those days. It was a thoroughly enjoyable trip with two of our dearest friends. Peter sat up with me until the early hours listening like the friend he was, to my continuous outpouring of the same emotions that aunty Sue had endured weeks earlier. I was still running away, from what I didn't know.

I was still confused, and I didn't have any sort of a plan, I knew that I was incapable of making major decisions and I knew that drink wasn't the answer, but what was?. I was still visiting dad in Ibstock, and we were still having fish lunches on a Saturday. I was on my way to fetch them and looking everywhere for faces that I knew, as I always did, but had not seen anyone familiar for years. Sue Tompkin as I knew her from way back, was talking to someone and I recognised her immediately. I carried on to the chippy and told of my sighting to Sue Taylor whom was serving. "I suppose it'll be another forty years before I see anyone I know in Ibstock," I said to her. "She'd be off to her mum's, she visits every day," she said. With that I left to enjoy lunch with dad.

Two weeks later, I was about to leave the house on Sunday morning, I hated Sundays at that time, and the phone rang. "Hello?" "Is that Donald?" I was taken aback, nobody calls me Donald any longer except my many aunties and the voice didn't resemble any of them. "Yes, speaking." "It's Sue Horner, just wondering how you are. I'm told that you saw me and asked after me a couple of weeks ago." Now it made sense, Sue had been a very good friend of my late sister Kate and was still a great friend of my aunt Winnie. I'd forgotten that her married name was Horner, after all, forty-four years had passed since I'd left the village. We chatted for a while and I found out that the grapevine had worked, and circumstances had connived for Sue Taylor to bump into Sue's son Scott at the swimming baths, she telling him of the sighting, he then telling his mum and Sue saying, "That's Kate's older brother and Winnie's nephew. I went to school with him." She then looked me up in the phone book, in which I had not long had our names put back and rang me a week later. I was thrilled that someone from so long ago had taken that trouble. "I'm in Ibstock again on Saturday, can I call and have a catch-up with you?" We arranged a time at her house, and I called after lunch at dad's. Both of us were a little nervous as the door opened to my knock but with tea and biscuits we relaxed a little as we sat to enjoy them. We chatted endlessly for around four hours after which I left, making similar arrangements for the following week.

I planned a golf break in Spain for a few days taking in three golf courses in the Fuengirola region, Stuart and Alun Davies jumping at the chance to come along on this December jolly. Sue and I had been out for a couple of meals together and were getting along well, although my mind was still all over the place and I told her so. The trip ended and had been successful in every respect and we still laugh about it today. I called Sue on arriving home and she said "I don't know how to tell you this, but your dad's been taken into hospital at the weekend. I didn't want to spoil your holiday by ringing you sooner." I thanked her and visited dad straight away, keeping everyone informed of his condition. We chatted at his bedside, "I'm glad that you've met someone Don, it was too late for me when I'd lost Peg, it was a lonely life without her." "Thanks dad." I was glad I'd told him of Sue and that I had his approval. Dad passed on Christmas day at 10:30 am.

RIP Dad.

Dad and I

'Born Lucky' Book Two: From Mynah to Miner and Beyond

Dad

'Born Lucky' Book Two: From Mynah to Miner and Beyond

New Beginnings

Dad's funeral was arranged and was soon, I panicked, Stuart and Andrea had not yet met Sue although they knew of her, I'd kept them abreast of our blossoming relationship. I forbade her from coming to Dad's funeral. "I'm not having your first meeting with my kids at Dad's funeral, it's not right." I told Andrea, she was appalled at what I'd done. "You can't do that Dad, she knew grandad and grandma and aunty Kate and is probably accompanying aunty Winnie." I knew I'd lost, but it didn't sit well with me at the time.

The service took place and I was amazed how many attended the funeral of a 92 years old man, but very proud at the same time. After the service, another mountain to climb, we were stood as a family Andrea, Carl, Stuart, Maria and I, and I'm sure they were as apprehensive as me. Various members of our family came to us and paid their respects and then Sue appeared with Winnie. Before I could say anything my aunty Sue rushed to me and hugged me and was very sad at her eldest brothers passing and we shared a few tears, the kids thought that that was the 'new' Sue. "Not at all, here's the Sue I want you to meet," and I introduced them individually.

That had gone better than expected, very mixed feelings as these things always are but to have to meet your father's new girlfriend at your grandad's funeral must be very daunting, but my family and Sue carried it off impeccably, as if it was an everyday occurrence. Sue and I grew closer, I was invited to her family gatherings and parties, Scott was amazing and I took to him immediately, we had a similar sense of humour. He was a Nottingham Forest supporter and with me being a Leicester fan, the verbal fencing was brilliant and still is. I met some old Ibstock friends too, Raymond 'Pet' and Dave Ralston with their wives, Stuart Farmer and Chris, Sue's brother but the loveliest couple were Jackie and Chris Woollerton whose son's party we were at. They were extremely welcoming and friendly and we were invited to their beautiful bungalow many times for a get together. We'd visit her friends and relatives, Pat Darby, Pat Ballard and 'Friday night' Pat, that's the trouble with having too many Pat's in your life, they have to be identified somehow. Another stand out introduction to me was Sue 'Tab' Taberner, a very funny lady and hairdresser, but more about her later.

After a while I suggested a holiday together. "I don't travel well Don, I've always suffered from travel sickness." "That's a shame, I love travelling." I told her of my biking trips and of my experiences around our country delivering cars. I convinced her that we should go, taking all of the preventative motion sickness remedies we could think of with us. Our first holiday together was to Zante (Zackinthos) and it went without a hitch, I kept a close eye on Sue and if I thought that there was something amiss, I'd talk to her, diverting her attention, but she was fine. The sun, the harbour, the vessels, the night sky and the wonderful cafes made for a great holiday that we both needed.

Golf and darts were still a big part of my life, Monday to Saturday was golf most days, Monday, Friday and Sunday evenings were darts with the occasional Thursday thrown in. I went on my first away trip with the 'C' team at a course in Shropshire, it was called the Masters weekend because it was arranged so that we could play on Sunday, have a presentation and then dinner and then watch the final round of the Masters, played at Augusta. USA. I travelled with Keith Tomblin and we played the initial round of golf on Sunday afternoon, watched the Masters etc. The Monday round was the singles competition, and I was drawn in a three ball with Keith and Bill Robson 'the hanging judge' as he was known, because his ruthlessness cutting the groups individual handicaps. We played the round, Bill marking my card, I really hadn't a clue what was a good score on a Stableford card. Was 41 in with a chance? Pete Rayers won runner up and announcing the winner Glenn Price (Captain) sai,d "Don Povey, where did you get your handicap from, a Christmas cracker?" I'd won but I was to be called 'Cracker' from that day forward and the hanging judge cut my unofficial handicap to thirteen. I had been dressed a bit garishly as I was prone to do and Keith would let me know about it and take me to a full-length mirror to make me see sense, it never did work. That day I had worn white shoes, grey checked trousers with blue and red lines running through, accompanied by a multi-coloured striped top and topped off with a blue peaked cap. I had also won a charity competition, nearest a bottle with a pound coin and so I decided that was what I was going to wear for the night's darts match.

Alun Davies, Stuart and I in Spain

John Butler picked me up as normal saying, "You're not going out like that are you? Miles (Hewitt) won't leave you alone." We arrived at the Crown and were playing Iggy's team, the scene was set for a good match, everyone was 'bobbying up' (practising) Alan Kearney had opened the free shots at me by calling me 'Rupert' as in the bear, everyone else chipping in with comments, I loved it and it gave me a chance to sharpen my wit with rebuffs. Miles came in through the door, he was sideways on to me whilst I was throwing. The room looked at him in anticipation knowing his character. "Only you would have the f**king nerve to wear that on a darts night." Everyone laughed. "Only you would insult a dwarf dressed for a charity do," was my answer to put Miles on the defensive. Everyone laughed again. Miles is at least 2m tall and a thoroughly good straight talking guy, if I asked what the weather was like up there, he'd look over my head, sniff and say, "I can smell a dwarf around here somewhere." We played the Anchor at Nether Broughton and there is a beam directly above the oche, bothering no one except Miles, even Scouse only just touched it with his hair, but Miles' forehead touched it when toeing the oche. He did no more than take off his size thirteen rigger boots and played in his socks. I played golf again on the Thursday with the same golf group, scoring 12 pars, 5 bogies and 1 double bogey, coming in with 43 points on a Stableford card with a score of 78, 7 over for the round. The hanging judge cut my handicap to nine after that and I wasn't even wearing outlandish clothes!.

I still had two cars on the drive in 2013, I hadn't the heart to sell my wife Sue's car, it seemed so final. My wife Sue and I had a discussion about me updating my Jag for a Jaguar XF, a new one. "Then get one," she said. "But I don't want to sell my S type, I think I'll keep it and get a new one anyway." "If you think I'm having two Jags on the drive as well as my car, you've another think coming and while you're at it, sell your motor bike too." I knew that would be the response that I'd get. I summoned up my inner strength ant ordered a new XF Premium Luxury 200. Stuart's reaction was, "I'm glad that you've done that dad, you deserve it." He claimed his mum's personalised plates SUZI POV and has transferred them to every car he's had since.

Sue and I were getting along famously, we would go away for weekends or just a night to lovely hotels near and far and just spend our time chatting, maybe a dance or just sitting. We had forty odd years to catch up on and it takes a while. I would ask, "Whatever happened to so and so?" "Oh he's dead," she'd answer and that happened a lot. "What have you been doing to them all since I've been gone? They're all dead or dying, my old friends." "It's not me," she'd say. Then I ask about someone else and she'd tell me their family history, their relatives, who they'd knocked off (unfaithful) and what ailments they'd got. I loved those conversations; it stretched my memory of people I'd known all those years ago and their relatives. We reminisced about our teenaged years and why our paths never crossed even though we lived within a very few hundred metres of each other for years, we recalled events and one of us would say "I was there that night too".

Holidays came around again, and we planned a Nile cruise, Sue being a little apprehensive about sailing but my reassurance worked and it was a great trip. We sailed from Luxor to Aswan, the boat was magnificent, and our suite was plush. The food, exquisite, the entertainment good and at times funny, the service to die for and our guide excellent. We marvelled at the temples, valley of the Kings and the impressive Karnack temple/town/villages. The Avenue of the Sphinx's when lit at night was a sight to behold, we loved that trip. Sue unfortunately had an upset tummy nearing the end of our two-week trip and I informed our guide. He was very concerned and jumped in his car, saying that their tummy bugs sometimes did not respond to British medication, "I'll go to the pharmacy and get you some of our medication," he said. He was gone for a while, returning with a small package. "Take one a day for two weeks and you'll be fine," he advised.

Sue and I on a cruise

Sue's bug didn't entirely clear up, so she made an appointment with her doctor, taking the remains of the package with her. "I've never heard of this before, but I'll look it up and let you know the result. In the meantime,e take this prescription and you should be OK," her doctor said and with that she left the surgery. A week later, Sue had cause to see a nurse. "Can you tell me what those tablets were I was given in Egypt nurse?" "I'll find out for you," the nurse answered and left. On returning, she said, "We don't prescribe those in this country, they're for animals." A shocked Sue said, "Bloody hell! I hope I haven't got a camel hump growing on my back have I?" Both of them laughing.

Weekend shopping trips were a regular event and trips to Burton upon Trent, Tamworth, Sheffield, Milton Keynes, Birmingham, McArthur Glen and Bicester were visited a number of times amongst others. Each time Sue would have to have a little treat as she'd refer to it. At first I thought it meant the coffee and cake we'd have, but learned later that it was the bags of clothes and shoes she'd bought every time! Sue even introduced me to her hair stylist, Sue Taberner, whom I've mentioned previously, and I went to have my hair cut at her shop on the High Street called 'A Cut Above'. Never have I had such an experience; it was amazing I've never felt so at home. Sue, Maria and Gemma should be a comedy trio, I laughed so much it hurt. Maria tells a story with all the animations and expressions; Sue backs her and eggs her on and Gemma chips in on things Maria has missed. Maria's kettle story is a must to hear. It's worth a visit for the show alone, I never tire of going there for a trim and a catch up with my favourite girls, Sue, Gemma and Maria.

Sue, Gemma and Maria

More holidays followed, Ibiza, Egypt and cruises, and I asked Sue to marry me on one Mediterranean cruise and she accepted. I had mentally planned it and had intended to carry out the quest the evening before after dinner but ducked out of it. The next evening I took a fake ring with me and after dinner at our table of eight I asked for their attention and asked Sue if she'd consider becoming my wife. Our cruise friends clapped, and the ladies came to Sue and congratulated her, as did friends we'd made from other tables. One said amongst the elation, "You haven't given Don your answer yet Sue, is it a 'yes' from you?" "Yes, I will," answered Sue and a cheer went up. The next evening the waiters all gathered around our table amid balloons and sang 'Congratulations' to us both.

Engaged

News of our engagement soon spread, and our families were all very pleased, Scott (Sue's son) was really pleased for his mum, Lleyton and Ryan, Sue's grandsons were equally pleased, and it was close to my 70th birthday and so we got together and had a family 'do' with Sue's and my families attending.

Sue, Gemma and Maria

About that time I was experiencing a slight problem with my left arm and leg, no pain just enough of a problem to know something wasn't quite right. I visited my GP Dr. Pearce Smith, who examined me thoroughly and said, "Come back in three months if it hasn't gone away." I re-visited him, and he referred me to a neurologist consultant Mr Ben Simpson. Symptoms at this stage were stiffness, tiredness and frequent falls although I was still enjoying a full life and no real problems. Test after test after test revealed nothing, Dat scans,

Mri Scans, GAD and AMG needle tests and still no answers, everything proving normal.

I was trying stuff on my own for the stiffness, Physiotherapy, Osteopaths, Hypnotists, Diets, gluten, dairy free and vegan, herbal remedies, acupuncture, extreme amounts of time exercising and swimming, Stuart even built me an overhead rail outside in the carport with a harness so that I could practice walking without the fear of falling, amongst other equipment he made for me. Of course I was spending huge amounts of time on my laptop researching symptoms long into the day and night looking all over the world, the most promising in Mexico, Australia and Pakistan but they revealed nothing. It's very frustrating to constantly find nothing and so I went to London for a second opinion. "They've been very thorough up there in Leicester," he remarked on seeing the notes I had provided him with. Mr Charles Kaplan examined me extensively with Stuart present and said, "If I were a betting man my money would be on Motor Neurone Disease (MND)." That was the first time it had been mentioned, it didn't bother me, at least it had a name, a handle, something that I could fight. "The trouble is that there is no test for MND, all the tests carried out eliminate everything else until only MND is left," he explained.

My darts left me first, I couldn't load from my left to right hand, I tried so hard I got dartitis, I got over that by throwing table tennis balls at the dartboard at home. Next to go was the 'feel' of the darts in my right hand resulting in over throwing or under throwing the target, and I couldn't even chalk properly with an outstretched arm, the guys knew something was amiss and I drifted from the spotlight to the sidelines. The golf then went, I lost 50m from my drive overnight, I was slow walking and occasionally falling until I had to stop. To my eternal shame I didn't tell the guys at the golf club that had helped me tremendously over Sue's passing. I did believe my condition was only temporary and that 'Cracker' would return, not to be unfortunately.

I applied to be part of numerous MND trials, to be turned down by them all 'Not the right age' or 'Not the type of MND we're studying' very disheartening, but we must remain positive. I was eventually accepted for the 'COMMEND' study conducted by the NHS. It took ten weeks of zoom calls, because of lockdown due to Covid 19, where I was interviewed for my thoughts on various subject matter and then sent a questionnaire by email to consider. Basically the end result was 'What don't you do now that you did before you were diagnosed with MND?' The answer being 'very little.' I can't play golf, darts, sing or walk but there's nothing else at present that I can't do and so I do it to lead a fuller life and in turn that lifts my moral and outlook. Friends and my very good neighbours will tell you I always say that I'm fine when asked, and I am.

I decided that I'd write a book. "My word Don, that's a big step to take and commitment," the psychologist, Roy McPartland said. "Maybe, but I'll do it." "Let me know how you get on with that Don, I'd be very interested in the result." I did let him know and he was amazed that I had carried out such a task, he bought a copy of 'BORN LUCKY but my life was THE PITS' that's the very reason I told him of it.

I've got slower, I don't talk so well but can make myself understood. I walk with a walker, and I visit Glenfield Hospital regularly, but I have a laugh with the nurses and Doctors. I still drive, type and think clearly and I'm very happy. I'm constantly told how well I look by my family and Sue's; I get first rate care and love from them all even though I still live on my own, I'm very very happy.

Born Lucky or what?

'Born Lucky' Book Two: From Mynah to Miner and Beyond

Sue and I

Roger, Di, Sue and I

Ruby and Megan my granddaughters

'Born Lucky' Book Two: From Mynah to Miner and Beyond

Scott and I

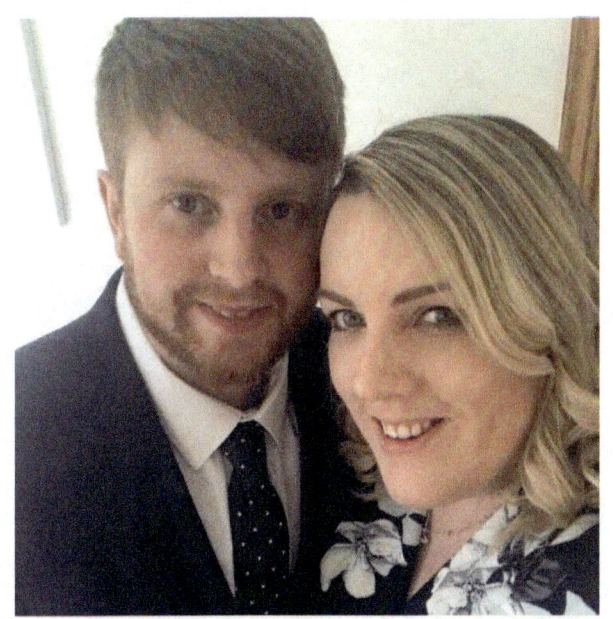

Grandkids Louis and Leanne

Grandson Jake

Sue's grandkids Ryan and Lleyton with her son Scott

Printed in Great Britain
by Amazon